Political Risk in the International Oil and Gas Industry

Political Risk in the International Oil and Gas Industry

Howard L. Lax

Atlantis, Inc.
Box 678
Green's Farms, CT

International Human Resources Development Corporation • Boston

Library of Congress Cataloging in Publication Data

Lax, Howard L.
    Political risk in the international oil and gas industry.

    Bibliography: p.
    Includes index.
    1. Petroleum industry and trade—United States. 2. Gas industry—United States. 3. Mineral industries—United States. 4. Investments, American. 5. International business enterprises—United States. 6. Risk.
    I. Title.
HD9565.L335   1983   622'.338'068   82-83329
ISBN 0-934634-20-3

For Susan,
without whom there is little

Contents

Foreword

The measurement of political risk has long been a part of commercial decision-making in the international petroleum industry. Events in recent years have broadened the audience for such skills, while at the same time called for even more careful assessments by political analysts in the industry. Consequently, it seemed this was the right time for a full-scale study of the contemporary techniques for political risk analysis as they applied to the global oil and gas business.

In this book, Howard Lax has drawn upon his training in political science and experience as an energy consultant at Atlantis, Inc. to focus on these issues. All of us on the staff have enjoyed contributing to this effort, and we would welcome reactions and comments from all readers.

W.G. Prast, Ph.D.
President
Atlantis, Inc.
February, 1983

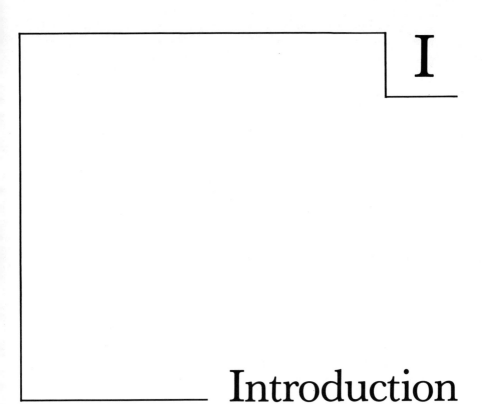

I

Introduction

1

Political Risk and Corporate Decision Making

Foreign Investment Decisions

The separation of economics and politics has been one of the key assumptions in the Western approach to the study of social sciences since the Industrial Revolution. Previously, economics was considered a branch of political philosophy or was ignored. The Greek *oikonomikos* referred to the management or rule over a household or estate. The mercantilist notion of the Renaissance era posited the acquisition of wealth (which actually meant filling the coffers of the King, the head of the "national household" or state) as an obligation of the state.

The era of contemporary economic thought was spawned by Adam Smith—who held a chair in moral philosophy, not economics—with the publication of *The Wealth of Nations* in 1776. Although the approach was a study in political economy and Smith acknowledged that politics and economics are inseparable, he and his followers drove a theoretical wedge between politics and economics. By positing "a natural economic order with laws of its own, independent of politics and functioning to the greatest profit of all concerned when political authority interfered least in its automatic operation,"[1] questions of economics and politics came to be seen as conceptually discrete.

The doctrines of eighteenth-century laissez-faire capitalism and nineteenth-century liberalism both assumed a separation between the respective domains of economics and politics and advocated a strict distinction as the most sound approach to solving international and domestic economic problems. Carried into the twentieth century, this classical approach viewed politics as an interference that introduced value conflicts and power concerns into the otherwise rigorously rational logic of economics.

The international economy was seen as the national economy writ large. This extrapolation usually was based on the examples of the United States or England. In making the leap from an individual country with a shared set of values and institutional processes to a far more heterogeneous world, theorists continued to assume a commonality based on "sound" economic reasoning. Discounting the political role of states in world affairs, economists spoke (and still speak) in terms of an economic rationality that assumes a harmony of interests that transcends national borders. Ignoring the political motivations implicit in such concepts as the national interest, solutions to economic problems were found in removing political influences from the international economy. It was at this higher plane of human affairs that economic questions were to be isolated from the unwarranted influence of international politics.

Although this idealistic approach has been modified greatly during the course of the century, the pragmatic capitalism that emerged in the post–World War II era still played down nonmarket influences in analyzing and formulating economics and business policies. Domestically, investment decisions were to be dictated by the anticipated rates of return from alternative business opportunities. Similar logic was applied to questions of foreign investment: investments should be made abroad when the expected foreign rate of return exceeded that offered on domestic investments.

In the "ideal" free market economy, companies no doubt would prefer conditions under which their operations were totally removed from political concerns. Increasingly, however, firms have come to realize that political variables constantly impinge upon economic issues that affect the company's operating environment. More immediate to a company's interests, political events and decisions often have a direct impact on even the most routine operations. Particularly in the international setting, the failure to treat economics as a field thoroughly interrelated with politics is a serious shortcoming in most modern economic thought.

In addition to traditional economic or business risks, corporate decision makers have begun to recognize a new category of risks that is political in nature. Foreign government interference in corporate operations was the exception rather than the rule throughout the 1950s and most of the 1960s. By the late 1960s, changing political conditions began to have a greater impact on the interests of firms with investments abroad, and the pattern of politically motivated changes in foreign operations became the norm during the 1970s. In the past five years, more than 60 percent of the U.S. firms operating abroad reported being subject to "politically inflicted damage."[2]

Although the domestic political environment affects company affairs, most of the focus on political risks deals with investment abroad. Companies tend to be more familiar and comfortable with the domestic environment. Executives understand the political processes at home, know how to influence political outcomes, and see the future as predictably stable. Foreign political conditions, on the other hand, are less well understood and therefore are seen as inherently more risky and less predictable.

Because each country constitutes a distinct business environment and political risks are usually identified with foreign investments, transnational corporations (TNCs)—firms that operate in a number of countries—are particularly concerned with the impact of political changes on corporate interests. Confronted with a multitude of operating environments, TNCs must cope with a variety of "political and economic systems, each with its attendant controls and risks, advantages and disadvantages, opportunities and dangers."³ Because each state has its own interests, traditions, and institutional processes, TNCs are subjected to a broad range of operating environments under a seemingly endless variety of conditions. The result is a proliferation of both traditional and political risks.

Having experienced repeated confrontations and negotiations with foreign governments and other political entities, TNCs have come to understand that the foreign investment relationship is inherently political. It is now recognized that transnational corporate operations cannot but be affected by and affect the political environment of the host country. The litany of forced divestitures, foreign exchange restrictions, unilateral agreement abrogations and changes, nationalizations, and expropriations,* to name but a few of the risks to which TNCs have been prey for the past two decades, provide mountains of evidence of how politics affects the operations of firms abroad.

Similarly, corporate activities, even if their motives are apolitical, affect political conditions in host countries. In the natural resource sector in particular, TNCs are essential to the performance of the national economy: many developing countries are almost totally dependent on earnings from exports of resources for hard currency and on tax receipts for government revenues. By the nature of their operations, the activities of TNCs are inseparable from issues of modernization, economic power, employment, and the distribution of wealth. Even in developed countries, the foreign control or ownership of natural resources—and their underlying value—is a sensitive political issue. Foreign firms also affect the foreign policy and international concerns of states in terms of such issues as trade, balance of payments, hard currency earnings, and the global distribution of wealth and consumption of resources.

Whether as a result of the exigencies of modernization in the Third World or of the demands of economic nationalism in the developed countries, governments are

---

*Although they are often used interchangeably, *nationalization* and *expropriation* are terms that refer to the same type of act but differ in scope. Expropriation is the seizure or modification of the property rights of a company or group of companies as an exercise of the sovereign power of the state, while nationalization is the same act aimed at a particular industry or sector of the economy.

trying to exercise greater control over foreign investors (and, in some instances, over domestic investors). In the most simple economic terms, state governments are trying to maximize the economic and political returns netted by the country and minimize the costs incurred through the operations of foreign companies. Conversely, TNCs are interested in maximizing their earnings from foreign operations and minimizing the additional cost burdens placed upon them by foreign governments.

The use of public policy to regulate sectors in an economy is not new, nor is the wont of governments to circumscribe the activities of foreign firms. What is new is that:

1. *the trend to greater government control involves almost all host countries, developed and developing, while in the past only a few of the more "radical" regimes were taking such actions;*
2. *the policies are becoming more explicit and detailed;*
3. *host government objectives have grown both in range and depth; and*
4. *host efforts are more likely to succeed than ever before because of the shift in bargaining strength (particularly with respect to natural resources) in favor of host countries.* [4]

The reality confronting firms going abroad is that political events influence operations more than ever before. The future promises to follow a similar pattern, by which political conditions will remain a central variable determining the viability of foreign investments and presenting TNCs with risks and opportunities.

Oil and Gas

To a large degree, host governments' assertiveness of control over their respective national economies has centered on the issue of natural resources. Petroleum—the most important resource, measured in terms of the value and tonnage of world trade,* its use as a material input in advanced economies, and the political intensity of concern on the part of host and home governments and TNCs—occupies center stage in the world political economy. As a logical correlation to the amount and intensity of political activity that has been focused on oil (the focus on natural gas has traditionally been low-key and only in the early 1980s has begun to assume international political proportions), the attendant political risks confronted by oil companies have been particularly high.

---

*The crude oil trade not only exceeds that of any other commodity in terms of both value and volume, it also is greater than that of any particular class of manufactured products.

Virtually every major oil-production agreement abroad has been subjected to renegotiation or unilateral alternation in favor of the host country. Foreign equity in crude has all but disappeared, exposing the oil firms to risks of supply shortages and interruptions and price instabilities. Although the nonconcession modes of agreement often imply a reduced capital investment on the part of the companies, which translates into a reduced amount of economic exposure to risk, the uncertainty and vagaries of the political economy of petroleum create a high-risk environment. Many of the modern service agreements, moreover, require that the oil companies assume the burden of the initial capital outlays involved in the preexploration stages of a project. As such expenses may involve tens of millions of dollars, the political risk that a host government will alter the terms of an agreement after a commercial find has been proved is very real to the companies involved.

All four of the previously mentioned trends are painfully evidenced in the world petroleum industy. Host governments of all political complexions, and states spanning the gamut from the Third World to the industrialized nations, have intervened to play a larger role in the development and disposition of their petroleum and natural gas resources. Democracy or autocracy, capitalism or socialism, North or South—when it comes to regulating foreign involvement in a state's petroleum and natural gas resources, countries tend to pursue similar goals of maximizing domestic economic returns and political control.

The explicitness and detail of host intentions are usually clear. Since the 1968 OPEC Declaratory Statement outlining ten particular policy goals, most petroleum producers have openly expressed their intentions. Often the policy aims with respect to the country's hydrocarbon development are incorporated into the general policy plans of the state. As a state's hydrocarbons policy has become increasingly important to broader domestic and foreign policy concerns, moreover, the range and type of government objectives with respect to petroleum and natural gas have expanded. Beyond the mere reflex action of maximizing economic returns, governments often seek to channel the petroleum sector to further multiple goals, including regional development, manpower training, welfare concerns, the distribution of wealth, political stability, and an array of other national priorities.

Finally, the measure of success that the oil-producing countries have enjoyed is evidenced by both the reactions of the consumer countries and the attempts at mimicry by countries whose economies are dependent on other primary resources. In economic terms, success is quantifiable in terms of the relative and absolute increases in petroleum prices,* the trade surpluses enjoyed and economic reserves

---

*Saudi marker crude, for example, was less than $2.50 per barrel a decade ago, compared to its designation at $34 per barrel in November 1981.

(mostly U.S. dollars) held by the more wealthy OPEC producers, the relative terms of trade between crude and other goods entering world trade, and the profit split between producer countries and oil companies. Although the current crude "surplus" (which has been partially engineered by Saudi Arabia) appears to have given oil companies and consumers some respite from price and supply pressures, the market in the 1980s is largely producer controlled. The success that oil-producing states have realized, moreover, has spilled over from the marketplace and has been translated into a wide range of political, social, military, and economic goals.

What is Political Risk?

Thus far we have not defined the concept of political risk. Like many terms in the social sciences, political risk is surrounded by a vast array of definitions, some better than others. We use the term *better* not in the normative sense of preference, but in reference to the descriptive and explanatory value of a definition.

By itself, risk commonly is used to refer to the "chance of injury, damage, or loss, compared with some previous standard."[5] Unlike uncertainty, which deals with a subjective potentiality of a loss, risks are measurable probabilities. Although the distinction between risk and uncertainty is often difficult to ascertain and may seem academic, there is a crucial conceptual difference: both terms refer to future likelihoods, but *risk* implies the ability to calculate probabilities and therefore to protect against and manage future contingencies, whereas *uncertainty* does not.

Risk is a dynamic concept that revolves around the probability of changes. Current conditions are not risks; rather, risk stems from *changes* in those conditions. The rules governing an investment are a current parameter at the time an investment decision is made. Risks are future occurences that may change the rules.

The adjective *political* carries a host of meanings. In its most narrow usage, it denotes the organizational and decison-making processes of governments. At its broadest, the term can be used to encompass virtually all the interactions between the units in a system (for example, people in a country or states in the international community). To avoid the pitfalls of being either encyclopedic or myopic in scope, we shall treat the term *political* as referring to the class of decisions and events that concern the authoritative allocation of values and resources[6] or that otherwise involve issues of legitimacy, authority, or the use of force.

In attempting to define political risk, analysts have shown a marked tendency to focus on such variables as sudden changes, instability, and government-initiated events. Although each of these variables may be a source of risks, they are too narrowly construed to serve as an adequate definition. Similarly, many definitions are too limited to incorporate nongovernmental activity or administrative procedures.

Conversely, there are those who use political risk as a "catchall term that refers to miscellaneous risks that are not otherwise known by particular names."[7] By throwing the net so broadly, this approach dilutes the utility of the term.

As an appropriate medium between the extremes, we employ a previously coined definition that still appears valid in that it accommodates a broad breadth of particular risks without being a "refuse bin" definition into which all inexplicable events can be classified:

> *In the generic sense, political risk is the probability that the goals of a project will be affected by changes in the political environment. It is the likelihood that political changes will prompt a change in the investment climate regulating a project.*[8]

In addition to the aforementioned advantages of our definition, it is attractive in that it accommodates a broad range of goals and focuses on the project involved. Although most corporate goals ultimately are reducible to the bottom line, the concept of profitability does not cover the full range of possible goals a company may pursue. By focusing on the project, moreover, we can address the political risks associated with nonequity agreements that technically are not classified as foreign direct investment. Firms that produce crude under production-sharing or other types of service agreements or explore for reserves under risk contracts are governed by the national investment regime and are subject to political risks.

On second reading, however, this definition is too limited in that it fails to satisfy the problems confronted by a company, for example, that purchases crude under contract. As only the broadest reading of the term *project* would incorporate contractual purchases, we prefer to amend the definition to "the probability that the goals of a project *or contractual agreement* will be affected by changes in the political environment."

A problem still arises regarding how to distinguish between political and "normal" commercial or business risks. Our approach is to focus on the "stimuli of potential change: does change stem from political acts and decisions on the part of government and other political actors or have market conditions changed without such influence?"[9] If changes are prompted by political acts and decisions (sometimes referred to as "risk events"), the risks are political. Changes in tax laws or in prices (they are government regulated) constitute political risks because, even though the result may be reduced profits, the source of the changes is political. National economic concerns (for example, an adverse balance-of-payments position or a shortage of foreign exchange) may be the ultimate cause of regulatory or statutory changes, but the decision to change is political and involves governmental processes.

Some firms prefer to use the term *environmental risks* or *constraints*. This introduces an undifferentiated mass of external influences and loses the value of focusing on those risks that are politically related. Although sociocultural matters may prove important to company operations, the culture of a society usually changes very slowly; moreover, when important changes do occur, they tend to be politically related.

Political risks, it must be borne in mind, imply the probability of both constraints and opportunities. Although risks often are considered only in a negative light (and certainly the focus of this book is on the adverse side of the coin), political risks also present opportunities. Tax incentives, price supports, and reduced regulatory restrictions are but a few examples of politically induced changes that present new opportunities for firms to further existing corporate goals and branch out toward new goals.

As a last word on how to think about political risk, one must be careful to focus on the "potentially significant managerial contingencies that result from political events or processes" and not the events themselves.[10] Political circumstances should be understood as the cause, with political risks as the corresponding effects. A firm needs to be concerned with how change in the political environment will affect that company's interests and not lose sight of the particular political risks, with their tangible results, by thinking in terms of more grandiose events that are distant and abstract in relationship to any particular company. This is not meant to imply that TNCs should ignore the larger picture and not keep track of national developments. To the contrary, the astute company will keep abreast of such developments but will interpret events in terms of how they affect that company and its interests.

Types of Political Risks

Political risks are not undifferentiated, homogeneous occurrences. The risks may vary between countries, between sectors of the economy, companies involved, and the particulars of each project. The types of risk to which a company is exposed vary accordingly.

Within a given economy, political risks can be dichotomized as micro or macro.[11] Macro political risks are aimed at all foreign firms, while micro risks are targeted toward particular sectors of the economy, types of firms, or even individual companies. Macro-level risks have a more sudden and dramatic impact on a state's economy and political risk profile, but micro risks are more common and have a more direct impact on foreign firms.

For our purposes, we classify types of risks in terms of the issues involved. Risks can be typed as transfer, operational, administrative/statutory, ownership, or contractual.[12] Transfer risks are concerned with the possibility of government restrictions with respect to the transfer abroad (and sometimes into the host country as well) of capital, profits, technology, personnel, equipment, or the actual commodity produced.

Operational risks entail the possibility that effective control over operations may be wrested from the foreign company and vested in the host government or its chosen representative or that government control may be used to the detriment of the company's interests. Operational risks are effective at the level of day-to-day operations and managerial control and the level of overall authority as represented by the board of directors or some equivalent body.

The issues classed as administrative/statutory include the likelihood that changes in the regulatory climate will affect a project or agreement. Ownership risks address questions of equity shares and involve issues of participation, expropriation, and nationalization. Whereas transfer, operational, and administrative/statutory risks involve both foreign direct investments and service agreements, ownership risks involve only direct investment projects.

Finally, contractual risks cover the bundle of supply/price issues embodied in nonequity, non-service-agreement transactions. Essentially, contractual political risks involve the issues of concern to companies that buy raw materials such as crude oil from foreign governments independent of any equity or service agreements between state and company.

Why Perform Risk Analysis?

Political risk analysis is not new. Any investment decision implicitly assumes a level of political risk. The traditional failure of firms to isolate political risk as a discrete variable to be considered in decision making does not mean that the risks were not present or that an evaluation, however rudimentary, was not made. A company's decision to pursue a project was ipso facto evidence that the firm considered the accompanying level of political risk acceptable. Refusal to enter into an agreement cannot, of course, be equated with a negative political judgment, since other concerns may have been the basis of the decision.

What is new is that firms increasingly have moved from implicit evaluation of political risks to explicit analysis of such risks. Confronted with multiple operating environments and continuously changing conditions in the world political economy that promise to affect the fortunes of TNCs, firms are recognizing that political conditions and anticipated changes are variables to be incorporated in their decision-making processes.

The trend toward political risk analysis is a logical progression in the pursuit of rational decision making by companies. A rational business decision based on anticipated rates of return, discounted cash flows, and alternative opportunities is impossible if an assessment of political risk is not included. Failure to measure risk precludes maximizing the rationality of decision making, as such risks have a vital bearing on the profitability and other goals of a project.

Political risks are present in any foreign investment or contractual agreement, especially when one party is a government or a governmental body. Ignorance of risks does not erase or minimize them; rather, it can leave the firm floundering in the seas of political uncertainty, unable even to estimate the degree of risk with which it is faced.

Risk analysis can move a company beyond the realm of uncertainty and enable it to systematically evaluate situations based on the analysis of information. By gathering information and subjecting it to analytical assessment, risk analysis moves a firm toward an explicit approach to evaluating risk. Because risks, as opposed to uncertainties, are measurable, can be taken into account in projected profitability, and can be managed, risk assessments are an integral part of educated and rational decision making.

By employing systematic, standardized risk analysis techniques, companies have been able to move beyond the superficial risk "feelings" that are often used as a justification for or reason to reject an investment opportunity. The traditional reliance on the intuition or experience of a seasoned manager yielded many insights. It is difficult, however, to incorporate insights into the decision-making formula, and the traditional approach prevented comparisons between foreign opportunities. Moreover, the tendency of many U.S.-trained business people to reject automatically any government's intervention in the marketplace as out of place and to interpret government's role strictly in terms of a potential setback to corporate goals interfered with efforts at objective risk assessment. The more impressionistic approach, furthermore, usually was based on a misplaced extrapolation of U.S. culture and values to a foreign country, and it tended to describe risks in general, sweeping terms that had little relevance to the specific interests of the particular firms involved.[13]

Confronted with dynamic political conditions that affect virtually all spheres of business activity, firms employ risk analysis as a decision-making tool. Risk assessment enables a TNC to understand better the behavior and roles of foreign governments, enabling the firm to improve its ability to manage its relationship with governments and their officials.

> *Profits can be made under a wide set of political, economic, social and cultural conditions. But the key to profits is planning the proper strategy* given prevailing conditions, *and the key to proper strategy formulation is* projecting conditions *with a sufficient degree of accuracy.*[14] *(emphasis in original)*

It is in the corporate planning function, therefore, that risk analysis is the most vital.

Petroleum and other natural resource firms and banks have been at the forefront of the corporate world in political risk analysis. This reflects the greater concern these companies and institutions have with political conditions and the sense of vulnerability that surrounds natural resource projects. Accordingly, in 1980 more than 70 percent of the TNCs in the natural resources field (petroleum and nonpetroleum) reported the use of "institutionalized assessment activities" that entailed explicit and systematic analysis of political risk.[15]

Forecasting Future Contingencies

Risk is a measure of the probability of changing conditions in the future. If a firm is entertaining an investment opportunity or contractual arrangement abroad, one can reasonably assume that the current political environment is at least acceptable and at best hospitable. Current conditions are known, allowing for relatively easy projections of rates of return and discounted cash flows based upon the prevailing fiscal regime of the country and the budgetary requirements and financial obligations of the project.

By definition, the future is inherently less certain. It is the future, however, that holds both the promise of profitability and the risk of changes that might reduce corporate returns or otherwise interfere with corporate interests. As decisions made today will be proved wise or foolish by consequences in the future, successful corporate decisions depend on an accurate projection of tomorrow.

The temporal difference between current and future political conditions is the dilemma. Time differences imply an important subtlety that, though ostensibly obvious, is often overlooked: political risk is not inherent in the current investment environment (although the roots of future risks may be); it is a result of possible changes in the environment. In other words, it is "changes in the rules themselves."[16] As such, a conceptional and policy distinction needs to be made by firms with respect to the current investment climate and the potential future climate.

Typically, most risk analyses are reactive, in that assessments are made in response to events in the firm or country involved. Upon first consideration of a project in country X, a company will evaluate the accompanying political risk. Rarely will an analysis be made in anticipation of or in advance of an expression of corporate interest in country X. Once an investment is made and a project has commenced, many firms update their risk forecasts for country X only in response to express changes in conditions that might affect the interests of the companies involved. In some instances, depending on the firm and the country, risk evaluations are done periodically, with a frequency ranging from six months to several years. Although changed conditions or events should prompt a new political risk evaluation, if assessments are limited to reactive functions, it reduces their utility in preparing the firm in advance for new contingencies. Anyone can wait until the future happens, but analysis after the fact will be too late for any valid planning purpose.

Political risk forecasts help companies plan and manage operations in anticipation of likely future contingencies. In any discipline, whether in the social sciences or the physical sciences, projections can be made confidently only on the basis of patterned and regular modes of behavior. Random events—such as the death of a head of state by natural causes or random acts of violence—cannot, by definition, be predicted. Most events, however, including assassinations and coups d'état, are planned and have root causes that can be understood, analyzed, and anticipated to some degree. The explicable and predictable, which constitute the overwhelming share of world events are understandable and essential ingredients of both accurate political risk evaluations and intelligent decision making.

The future needs to be understood not merely in terms of general politico-economic trends and a vague sense of a country's attitude toward foreign firms, but in the sense of what the future means for a particular company (given its country of domicile, sector of operations, and approach to management). For political risk to have any substantive meaning to a particular firm, risk must be understood in terms of future managerial contingencies that may arise and how such contingencies will affect corporate goals. Having asked the preliminary question, "What is the political environment likely to look like?" the most useful analysis will be cast in terms of "How will that affect my firm?"[17]

The Decision-Making Process

For a political risk assessment to be of any use, it not only must be thoroughly prepared, but it must be used by senior management. Although this may seem an obvious truism, it points to the crucial question of how political risk analyses are used in decision making.

Whether decision makers are receptive to political risk assessments depends on their predisposition toward political questions (that is, whether or not they are interested) and the organizational structure of the firm. We will return to base issues later, but it is important to recognize here that some decision makers do not (and do not want to) consider political developments sufficiently important to merit much attention. There is a larger group that is concerned with political circumstances but does not trust the validity of political risk analyses or does not consider the information to be in a usable form.

On the whole, political risk is recognized as an important variable in the corporate decision-making process. Nonetheless, *how* important risk forecasts are often is contingent upon corporate organization. As an inherently interdisciplinary, non-traditional field, political risk analysis does not fit neatly into the regular corporate niches. Often buried in the corporate planning or financial division as the part-time job of a junior person, risk analysis lacked sufficient priority and access to decision makers to make any impact. In moving away from the traditional mold in risk forecasting, many firms have begun to employ specialists in the field (usually political scientists or ex-foreign service officers) and to organize the assessment function as a quasi-independent unit, often under a senior person who has direct access to top management.

In addition to the internally generated data and political risk evaluations, many companies use outside risk assessment services. The growing importance of risk analysis is evidenced by the growth in the number of consulting firms that provide risk evaluatory services The activities of these firms range from providing highly specialized services to preparing general surveys of dozens of countries. A number of older and larger research and consulting firms are active in this field, along with a host of smaller newcomers. Although many TNCs subscribe to or use the specialized services of risk assessment firms, there is a general consensus that such external analyses are less likely to be used by corporate management than are internally produced assessments.

A number of ex-government officials also provide political risk consulting services. Among the better known names, three former CIA directors—William Colby, Richard Helms, and James Schlesinger—are in the risk advisory business, and former secretary of state and national security advisor Henry Kissinger is a political advisor to Chase Manhattan Bank. Political risk services from former officials such as these tend to be the almost exclusive domain of the largest corporations. Many other ex-government officials of lesser stature also offer their services as political risk analysts.

Utility of Political Risk Analysis

The importance of political events and decisions to the operations of foreign firms has been punctuated by the inability of apolitical economic models to predict outcomes.[18] Predictions that OPEC would disintegrate under the pressures of intermember competition for market outlets[19] or that the crude price that "makes the most sense" is the one that maximizes the discounted cash flow to each oil-producing country[20] have been based strictly on economic theory—and they have been wrong. Despite the best efforts to study risk solely in the light of the economic theory of a freely competitive marketplace, the world refuses to behave according to economists' expectations. The realities of an international system comprising more than 150 sovereign states, each of which has its own interests, militate against use of the economic model as an adequate explanation (and prediction) of state behavior. Similarly, although domestic political life (at least in the more stable countries) usually is more orderly than international relations, a state's hierarchy of power, interests, and policies is political by nature and rarely will be in agreement with an abstract determination of an economic utility function.

Rational decision making assumes full utilization of information and a reasonable ability to estimate the likelihood of alternative outcomes. Insofar as this purpose is served by adding the political dimension to decision-making processes, political risk analysis is an important part of rationalizing decisions. Risk assessments both reduce the margin of uncertainty and address a spectrum of possible political developments that are important to the operations of a firm yet are beyond the focus of economic theory.

All decisions entail value tradeoffs: because of capital and manpower concerns, exploration commitments in one country may preclude similar activities elsewhere; pipeline investments might prevent a firm from adding to its tanker fleet; involvement in an LNG project might reduce a company's ability to diversify its natural gas sources. When confronted with alternative opportunities, decision makers must evaluate the relative costs and benefits of the opportunities before a rational decision can be made. Political risk analyses add to this goal in two ways. First, risk assessment techniques enable the firm to minimize the degree of uncertainty inherent in all situations. Second, systematic risk evaluation allows comparisons between countries and opportunities. It makes sense to measure the bullishness of a country in its policies toward pricing crude or its proclivity to change the fiscal regime governing the production and sale of hydrocarbons only in comparison to another country or a common standard. The commonly used adjectives with respect to OPEC countries—moderate and radical—have no meaning except in comparison with other countries that are more or less so.

Failure to employ political risk analysis leaves a firm prey to the polar risks of passing up lucrative opportunities and getting involved in costly, unrewarding projects. This is particularly true when decision makers rely on general, sweeping impressions of countries. Contrary to what might be expected intuitively, an oil investment in Angola may outlast and prove more profitable than a project in Canada. The "improved capacity to assess overseas political environments may provide a distinct competitive advantage in indicating where host country leaders are prepared to offer new incentives, to bargain among chosen objectives, or otherwise enhance business projects."[21]

Not only does political risk analysis permit comparisons between countries, it also affords insightful comparisons of a given country's political environment over time. Similarly, a country's political risk profile can be used to frame the requisite investment and contractual agreements or to recommend the incentives that are necessary to make a project attractive. Analytical approaches to political risk enable decision makers to see beyond the events that are popularized as catastrophic by the media and to minimize the influences of the "odd-lot syndrome"—the tendency to focus on and overemphasize the importance of particularly dramatic events.[22] Failure to be able to measure value tradeoffs and compare opportunities, particularly in a time when political variables are growing in importance, indicates that greater costs may be incurred by *not* analyzing political risk than by employing the resources necessary for professional analysis.

Although it would be improper to claim that corporations maintain a "foreign policy" toward host states, all firms have what we shall refer to as external policies. Such policies are directed toward both host and home governments, although the content is likely to differ. External policies (as distinct from internal policies) refer to the patterns of behavior and the principles a firm employs in dealing with others, including states. To structure a logical, coherent external policy abroad—whether for reasons of negotiation, to be a "good corporate citizen," or to learn about the local power structure—a firm needs to understand the political environment. This is particularly important to highly visible firms in crucial economic sectors, since, by virtue of the role they play in a state's economy, they become part of a country's political environment.

Risks are manageable. Management of political risks, however, presupposes a knowledge and understanding of such risks. As it becomes increasingly apparent that the breadth of options available to host governments is greater than ever before and that hosts are more likely than ever to be successful in a direct conflict with a foreign company, TNCs need to be prepared for possible future contingencies. In contrast to host states, TNCs seem to be more limited than previously in their ability to press their demands on governments successfully. Political risk assessment provides corporate managers with an insight into the subtleties of the political environment, better preparing them to be alert to potential changes and to deal successfully with government officials.

In summary, political risk analysis serves three vital functions: it enables a company to approach rationally the question of whether a particular investment project or contractual agreement should be pursued; it keeps a firm apprised of political conditions and their impact on corporate interests, thereby permitting the intelligent management of political risk; and it helps a firm in articulating an external policy that is appropriate to the particular national context. In every instance, the ultimate measure of political risk analysis is its ability to forecast future political developments and to explain how such developments will affect the commercial interests of a foreign firm.

Notes

1. Edward H. Carr, *The Twenty Years' Crisis* (New York: Harper and Row, 1964), p. 114.
2. *Wall Street Journal*, 30 March 1981.
3. Michael Z. Brooke and H. Lee Remmers, *The Strategy of Multinational Enterprise: Organization and Finance* (London: Longman Group, 1970), p. 9.
4. C. Fred Bergsten, "Coming Investment Wars?," *Foreign Affairs* 53 (October 1974): 136–137.
5. Yair Aharoni, *The Foreign Investment Decision Process* (Cambridge: Harvard University Press, 1966), p. 37.
6. David Easton, *The Political System* (New York: Knopf, 1953).
7. Thomas L. Brewer, "Political Risk Assessment for Foreign Direct Investment Decisions: Better Methods for Better Results," *Columbia Journal of World Business* 9 (Spring 1981): 5.
8. W.G. Prast and Howard L. Lax, "Political Risk as a Variable in TNC Decision Making," *Natural Resources Forum* 6 (April 1982): p. 185.
9. Ibid.
10. Stephen J. Kobrin, "Political Assessment by International Firms: Models or Methodologies," *Journal of Policy Modeling* 3 (1981): 253.

11. Stefan H. Robock, "Political Risk: Identification and Assessment," *Columbia Journal of World Business* 11 (July–August 1971): 9–10.

12. This typology builds upon that presented in Franklin R. Root, "Analyzing Political Risks in International Business," in *The Multinational Enterprise in Transition*, ed. A. Kapoor and Phillip D. Grub (Princeton: Darwin Press, 1972), p. 357.

13. Aharoni, *Foreign Investment Decision*, p. 94.

14. Business International research report. *Strategic Planning for International Corporations* (New York: Business International Corporation, 1979), pp. 73–74.

15. Stephen Blank, *Assessing the Political Environment: An Emerging Function in International Companies* (New York: Conference Board, 1980), p. 8.

16. Lars H. Thunell, *Political Risks in International Business: Investment Behavior of Multinational Corporations* (New York: Praeger, 1977), p. 5.

17. Stephen J. Kobrin, "When Does Political Instability Result in Increased Investment Risk?," *Columbia Journal of World Business* 13 (Fall 1978): 121.

18. This argument is presented in William D. Coplin and Michael K. O'Leary, "Political Analysis in the Forecast of Oil Price and Production Decision," paper presented at the Electric Power Research Institute Fuel Supply Seminar, Memphis, 8–10 December 1981, pp. 1–9.

19. See Morris Adelman, *The World Petroleum Market* (Baltimore: Johns Hopkins University Press, 1972).

20. Robert S. Pindyck, "OPEC's Threat to the West," *Foreign Policy* 30 (Spring 1978): 37.

21. Blank, *Assessing the Political Environment*, p. 3.

22. R.J. Rummel and David A. Heenan, "How Multinationals Analyze Political Risk," *Harvard Business Review* 56 (January–February 1978): 68.

II

The International Petroleum Industry

2

The Politicization of Oil and Gas

The Traditional Relationship Between Oil Companies and Host Governments

The history of private foreign investment in petroleum dates back to the turn of the century. European interests, particularly British and German concerns, were anxious to gain access to the promising oil reserves of the Middle East. The first petroleum "concession"—the term used to describe the traditional agreements under which foreign firms had extensive and exclusive rights to exploit the natural resources of a specified area—was granted by the shah of Persia to William D'Arcy in 1901. This concession was assumed by British Petroleum in 1909.

The scramble by the European powers to win concessions in the Middle East was mirrored by the activities of American companies in Mexico and Latin America, Russians in Rumania, the Dutch in the East Indies, and British firms in India and Burma. World War I both interrupted the concession acquisition process and emphasized the strategic importance of petroleum. Following the close of the war, American oil interests, supported by the U.S. Department of State, attempted to get a foothold in Middle East Oil, and France asserted its claims to the concession rights previously held by Germany.

The interwar period saw many petroleum concessions granted, as virtually all of the known and suspected oil-rich areas of the world were divided into territorial assignments held by the international petroleum companies. Concessions covering

Saudi Arabia, Iran, Iraq, Bahrain, Kuwait, Abu Dhabi, and Qatar were granted (or modified to accommodate different participants) during these years. The eight largest oil firms—the international "majors" as they came to be known—Exxon, Gulf, Texaco, Mobil, Standard of California (Socal), British Petroleum (BP), Royal Dutch Shell, and Compagnie Française de Petroles (CFP)—controlled between 95 and 100 percent of each of the Middle East holdings.*

The legal relationship between the oil companies and the host governments in the oil countries was governed by the terms of these agreements. Under the traditional arrangements, the concessionaires had exclusive and plenary rights covering all aspects of petroleum-related operations and activities. The general conditions were largely standardized and included:

— extensive areas (the D'Arcy assignment covered 500,000 square miles, five-sixths of modern-day Iran), with no relinquishment requirements;
— lengthy concession periods (the Socal concession for Saudi Arabia was for 66 years);
— 100 percent equity investments by the foreign firms, which had ownership rights to the petroleum produced;
— initial cash payments (Ibn Saud got £50,000 in gold upon granting the Saudi concession), supplemented by fixed royalties based on volume (there were no income taxes); and
— total managerial control over operations and planning vested in the concessionaire.

The political relationship behind the agreements was analogous to that of colony and colonizer, small territory and great power. The Middle East was more an amalgam of fiercely independent, divided sheikdoms than a region of sovereign states. The dissolution of the remains of the Ottoman Empire in the course of World War I ended the few vestiges of political coordination that existed. After the war, political responsibility and control was assumed by the Europeans, particularly Britain.

Economically, oil was the link between the Middle East and the world market. The demand centers for petroleum were in Europe and the United States (which had sufficient, albeit more costly, domestic petroleum reserves). This demand gave Middle Eastern oil its value. The local economies of the area had little use for the oil. Without the European and American markets, the crude would have remained beneath the desert sands.

---

*At the time, Exxon was known as Standard Oil (NJ), Mobil as Standard Oil Co. of New York (Socony) and then Socony-Vacuum, and BP as Anglo-Persian and then Anglo-Iranian Oil Company.

The relationship between consumer and producer countries institutionalized a flow of trade and a symbiotic relationship that netted inexpensive crude for the companies and previously undreamed riches for the local rulers (in exchange for a commodity that had no value other than for export. The international oil companies had the technology, capital, skilled labor, and market outlets to develop Middle Eastern crude oil resources. Lacking the vital inputs necessary to produce oil, having virtually no domestic demand, and being politically and economically unable to mount an effective challenge to the concession relationship, the Middle Eastern rulers had little negotiating strength in their dealings with the oil companies.

The oil firms did have to indulge the leaders on whose authority concessions were granted. Additional cash payments and gifts were not uncommon. Ibn Saud, in particular, always seemed to be demanding more—and getting it—from the Saudi Arabia concessionaires. As was typical of the premodern, quasi-independent countries of the time, however, local rulers exercised virtual ownership rights over their domains and substituted the maximization of personal wealth for any social welfare function. Satisfying the sometimes exorbitant tastes of local rulers was a small price to pay in exchange for access to and control over the area's petroleum reserves.

Resources and Development

Development is a less-than-specific concept for describing the growth process and goals of states. The distinctions between *developing* and *developed* countries comprise an array of relative and absolute measures. In contrast to the developed world, developing nations as a group tend to have less diversified, largely preindustrial economies. Subsistence agriculture and single exports often are the backbone of the national economy. Other distinguishing characteristics are lower standards of living (measured in terms of per capita GNP, infant mortality, life expectancy, consumption of raw materials, literacy, and other socioeconomic indicators), a less-developed technical and social infrastructure (fewer and poorer quality communications, transportation, and social welfare systems), and less-institutionalized central political systems (poorly trained and staffed bureaucracies; local leaders and norms often prevailing over national authorities).

The adjective *developing*, used to describe the poorer countries, implies both a process and a goal: to develop is to improve a country's socioeconomic condition, to have a higher standard of living, and to be part of the modern world. The Western countries developed over the course of two centuries or more, sharing common values and having extensive untapped markets into which they were able to expand. Third World nations face a compressed time frame; the need to develop is a chronic and immediate problem for which there are only uncertain solutions.

Drawing on the experiences of such Western resource-rich countries as the United States, Canada, and Australia, the traditional view is that the developing countries should rely on the export of primary commodities as their vehicle of development. The vital inputs that are not available locally can be obtained from foreign investors, in exchange for which the minerals-producing countries can export natural resources.

This school of thought began to come under attack in the 1950s by Third World leaders and academicians who were dissatisfied with the limited progress their countries were making. Led by articulate spokesmen such as Raul Pebisch,—the first head of the United Nations Economic Commission for Latin America (ECLA) and first secretary-general of the United Nations Conference on Trade and Development (UNCTAD),—and representing the interests of the emerging numbers of independent Third World countries, the alternative view claimed that the producer/consumer relationship meant permanent "underdevelopment" for developing countries. The terms of trade (the relative market value of commodities and goods traded) and the world political economy, the argument ran, systematically penalized the raw materials exporters. The gains from foreign investment were questioned, and the benefits of Third World minerals were seen as flowing to the developed countries while development efforts in the poorer countries floundered.

Without extractive industry wealth to attract foreign capital and to provide exports, many developing countries might not have made any gains in moving beyond subsistence agriculture. Simultaneously, however, mining and petroleum industries tended to be developed as enclaves that were not integrated into the larger national economy. Dual economies emerged as large, modern natural resource projects existed side by side with the traditional agricultural sector.

It became apparent that the exploitation of a country's natural resources was an important element in approaching the broader issue of development. Technology, capital, and skills—all vital inputs in the developing process—were obtained only through the transnational firms. For host government leaders to maximize the development benefits realized from an oil or mineral project, however, it was necessary to challenge the authority traditionally exercised by the foreign investors.

The logic of development has contributed to host government efforts to seize control over resource-related projects. The previously mentioned pattern of resource trade between the developed and developing countries has made raw material commodities (and related products) the leading export of the Third World. In terms of value and volume, crude oil is the most important commodity in world trade. As a result, a very large share of export earnings and hard currency receipts is determined by the volume and price of oil and minerals exports. An even larger share of host government revenues is dependent on natural resources. Saudi Arabia, Libya, Iraq, and Kuwait, for example, rely on oil sales for between 90 and 99 percent of all export earnings and for virtually all government revenues. Even a relatively advanced and diversified developing country such as Venezuela is dependent on petroleum exports for approximately two-thirds of its government revenues.

Because oil and other resources are so important to the income of the resource-exporting countries, control over resource development has been a prime political target for a number of years. To a large extent, political risks associated with deliberate governmental changes in policy are a result of the importance petroleum plays in the economies of the oil-exporting countries. In the Middle East, planning of petroleum policies is tantamount to economic development.

It is not likely that a government's interests in channeling petroleum projects to further the goals of national development will be shared fully by an oil company. Company interests, logically, are in making profits, remitting profits abroad, and obtaining supply. Host governments, however, have their own profit maximization goals, which frequently are not coincident with those of the oil firms. Aside from maximizing profits, moreover, oil-exporting countries are interested in directing oil projects to serve domestic policy goals. These goals may include such items as promoting regional development, improving the distribution of wealth, keeping the domestic prices of petroleum products below world market levels, employing and training nationals, and building a national oil company. Oil firms have little or no motive to share these goals, except insofar as the bargaining leverage of host governments has forced the firms to address their attention to these issues. As this has happened, the companies have been compelled to shoulder more costs. The conflicting interests of host and firm will be explored later in this chapter.

The Changing Relationship

The concession relationship largely insulated the oil companies from political interference by the host government leaders. Since the original agreements were signed, however, the firms have been under constant pressure to offer better, more rewarding terms to the host countries.

Over time, there has been a steady process by which the terms of project agreements have increasingly favored the host states. Bargaining power has steadily shifted toward the host governments, which have used their bargaining strength to increase revenues and managerial controls.

Even before World War II, changes were under way in the structure of the world oil industry and in the concession relationship. Host governments began to insist on shorter time periods, smaller areas, and scheduled relinquishment. More important to the future economic aspirations of the oil-producing countries, Venezuela's 1943 Law of Hydrocarbons unilaterally canceled the existing concessions and imposed a 16⅔ percent royalty, based on the market value of oil (as opposed to volume), on new concessions.

Following the war, dozens of independent U.S. and European oil firms began to compete for oil exploration rights abroad. The first important concessions won by independents covered the Kuwait and Saudi neutral zones and were granted in 1948 to Aminoil and 1949 to Getty, respectively.

At the same time, Venezuela introduced income taxes for oil-producing firms. In 1948, Venezuela instituted a 50 percent tax on all profits. The idea of 50–50 profit sharing was adopted by Saudi Arabia in 1950 and became standardized throughout the region in the early 1950s.* Despite the inroads made into company per-barrel profits, the host countries were not challenging the firms' control over concessions or ownership rights to production.

The Iranian nationalization was the first major political challenge for the oil companies in the Middle East. A united front by the major firms (assisted by support from the U.S. and British governments) frustrated the nationalization program of the Mossadegh regime. The settlement, however, introduced a number of independent companies to the new Iranian Consortium, planting the seed for future internecine warfare between the integrated and independent oil firms. Moreover, Iranian oil remained nationalized. Although it may have been only a legal distinction, since the Consortium operated essentially the same as BP did prior to 1951, the concessionaires no longer owned the oil at the point of extraction. In the process of nationalization, the first Middle Eastern national oil company—the National Iranian Oil Company (NIOC)—was born.

Long a leader among the national oil companies, NIOC negotiated the first joint venture agreements in 1957 and 1958—the former with Ente Nazionale Idrocarburi (ENI), the Italian state-owned company, and the latter with a subsidiary of Standard Oil of Indiana. The NIOC joint ventures were the first time that a host government had an equity share in operations with a foreign oil company.

In the mid-1960s, Indonesia pioneered the production-sharing agreement. The Indonesian deals recognized the state's ownership of the oil produced and provided for a sharing of production rather than profits. In 1967 a production-sharing agreement was reached with Conoco—the first such agreement with an international oil company.

In 1966, the same year as the first Indonesian production-sharing initiative, Iran took another innovative step, entering into a service contract with Enterprise de Recherches et d'Activités Petrolières (ERAP), a French firm. The agreement area comprised acreage relinquished to the state by the Consortium. Although the project met with limited success, it marked the introduction of what is now an important means by which the oil producers exercise direct control over petroleum operations.

---

*Because all royalties and fees paid to the government were credited against the companies' tax burden, the countries received less than 50 percent of the profits. It was not until royalties were expensed in the 1960s (that is, treated as tax deductions rather than as credits) that the profit split actually was 50–50.

The Emergence of OPEC

The first hints of host government cooperation reach back to 1949, when a Venezuelan delegation led by Petroleum Minister Juan Perez Alfonso went to Iran, Iraq, Kuwait, and Saudi Arabia to lobby for and explain the advantages of adopting a 50–50 profit arrangement. The first formal agreement, signed by Saudi Arabia and Iraq in 1953, provided for the exchange of information and ideas about the petroleum industry. Venezuela's Alfonso repeatedly called for closer cooperation between the countries as a means of furthering their common and respective interests.

The Organization of Petroleum Exporting Countries (OPEC) was formed in 1960 as a reaction against the second cut in posted prices by the international companies in two years.* The attention given to the new organization reflected the impact the Western states and firms expected it to have on the world petroleum industry: the *New York Times* covered the inaugural meeting with a filler-type piece buried deep in the paper.

Befitting its power at the time, OPEC's initial goals were modest: restoration of prices to the levels existing before the reductions and consultation by the companies with the countries about price changes. During its first few years, the organization groped about, uncertain of its goals or power, obtaining minor improvements in terms from the oil companies.

OPEC unveiled its grand design for gaining control of the industry in its "Declaratory Statement of Petroleum Policy in Member Countries," adopted at the organization's conference in 1968. The Declaratory Statement highlighted many of the principles that subsequently guided the policies adopted by OPEC members and non-OPEC oil producers. The Declaratory Statement recommended the following policies:[1]

— to the maximum extent feasible, member states should explore for and develop their oil and gas resources directly;
— to the extent that foreign assistance is needed, the government should retain a maximum share in the equity of and control over all projects;
— based on the principle of changing circumstances, governments "may acquire a reasonable participation" in current concessions and operations;
— a program of progressive and accelerated relinquishment should be instituted;
— taxes and payments to states should be based on posted prices to be determined by the government, with the prices of the various member governments consistent, given differences of geography and quality;
— "excessively high net earnings" are grounds for renegotiation; and
— disputes will be settled by the appropriate national or regional courts.

---

*Posted prices are those on which taxes and royalties were based. Selling prices could differ.

At the time, it appeared that OPEC had set its aims too high. The Declaratory Statement was calling for the restructuring of the world industry in accordance with the demands of the host countries. The process by which the companies were continuously defending against the erosion of their control and market power, however, had already gained momentum.

A number of governments were able to use their political prerogatives to their advantage in dealing with the companies. During the 1960s, in addition to the production-sharing agreements in Indonesia and the service contracts in Iran, the oil companies were subject to repeated nationalizations, the most important of which was the Iraqi seizure of 99.5 percent of the Iraq Petroleum Company concession (Law No. 80, 1961). Distribution and refinery operations, along with some production assets, were subject to nationalization in Ceylon (1962 and 1967), the United Arab Republic (1964), Syria (1965), and Algeria (1965 and 1967). In the Western hemisphere, Peru nationalized the International Petroleum Company (IPC) in 1968, and Bolivia seized the assets of Gulf the following year.

The emerging power of the producer countries was apparent in other areas as well. As far back as 1960, Shell was the first major company to accept limited government participation in an offshore project with Kuwait. National oil companies, the future instruments of host government policy, were being established during this period:

— Corporacion Venezolana del Petroleo (CVP), 1960;
— Kuwait National Petroleum Company (KNPC), 1960;
— General Petroleum and Mineral Organization (Petromin, Saudi Arabia), 1962;
— Société Nationale pour le Transport et la Commercialisation des Hydrocarbures (Sonatrach, Algeria), 1963;
— Iraq National Oil Company (INOC), 1964; and
— several Indonesian state agencies, notably Pertamina.

Moreover, agreements in Indonesia, Iran, and Libya (which commenced production in 1961) were modified in favor of the host countries.

It almost seems that, in the rapid pace of events, the political and economic power accumulated by the producer states went unnoticed. Even the leaders of the OPEC nations seemed unaware that the industry's structure of power and control was shifting in their favor. The more the Western world consumed oil and relied on imports, the more the independent oil companies flocked to compete for the economic returns promised by the oil industry. The more the major oil companies and consumer states relied on OPEC oil, the more the Western states and firms unknowingly contributed to the growth of OPEC oil power—and that of subsequent non-OPEC oil exporters.

The tide of events in the early 1970s—the Libyan showdown with Occidental in 1970, the agreements at Tehran and Tripoli in ·1971, the Geneva agreement later that year, the participation agreements of 1972, and the embargo and unilateral price increases of 1973–1974—caught many political and business leaders off guard as, in the course of a decade, control over an entire industry seemed to change hands.

The Politics of Change

The extractive industries—particularly petroleum—have long been the most sensitive area of foreign involvement in the Third World, the cutting edge of the bundle of issues separating the North from the South. Because of the visibility of petroleum operations, the importance of oil wealth and income to host economies, the vital role afforded to petroleum policies as the key to development, and the legacy of Western economic involvement in developing countries, the oil industry has proved to be especially sensitive to the turn of political events. Conversely, the absolute and relative importance of oil in world affairs makes changes in the international industry politically significant.

Over time, petroleum and mineral agreements have been subject to what one analyst has termed "political obsolescence . . . the deterioration in profitability of an investment because of political rather than other reasons."[2] Investment agreements are the product of an ongoing bargaining relationship between host governments and international oil companies. The outcome of negotiations, which is always the starting point for a fresh bout of tacit or explicit bargaining, is determined by the relative bargaining strengths or power each party can bring to bear on the process. "The bargaining process, moreover, does not occur in a vacuum; rather it takes place in a fluid, changing international environment of ongoing political and economic activities."[3]

The bargaining power of the oil firms was greatest when they enjoyed unchallenged control over the vital inputs—capital and skills—needed to search for, produce, and market oil. As that control waned, the firms experienced an erosion of their bargaining position. The availability of capital from other sources, the large number of independent oil companies vying for access to crude, the existence of highly specialized companies (which often are used by the oil firms as well as by host governments) that provide engineering and other services needed for petroleum development projects for fees, and the emergence of national oil companies all have contributed to the decline in the market position of and bargaining power wielded by the oil companies.

With respect to a particular agreement or relationship between a host state and a foreign oil firm, the company's position is strongest at the outset. It is at this point that the country is most dependent on the firm to supply the missing or scarce factors of production.[4] At the preagreement stage, moreover, the economic and geological risks are highest, often making it unacceptable for the host government to commit national funds for exploration. If a company controls the vital production inputs a country needs—a package or bundle of factors of production or a single essential input—it will enjoy a measure of leverage during negotiations. Before committing itself to a project, furthermore, the firm incurs no risks except the opportunity risk.

As a project proceeds, however, there is a tendency for the balance of bargaining strengths to shift steadily in favor of the host government. Once a firm has invested in a project, it has a tangible stake and risk in the outcome. After exploration is completed, the economic risks associated with a specific project are reduced sharply. This decreases the leverage of the firm during the postexploration stages. Similarly, after construction is complete and production has begun, the relative bargaining strength of the oil company tends to slip even more.

During the course of natural resource projects, the host governments climb the learning curve. The more indigenous capabilities and skills the national oil companies and host governments nurture, the less vital is the foreign firm. With large fixed-capital investments in place and reliance on a continuous source of supply, the firm becomes increasingly vulnerable to political risk; it has that much less bargaining leverage and that much more of a tangible interest to protect. The host state, on the other hand, develops a cadre of educated and skilled leaders and negotiators. The government learns about the industry and develops a new competence in negotiating and dealing with the large transnational firms.

In a simplified description, the bargaining relationship evolves along a continuum along which the bargaining advantage initially rests with the oil company but continuously moves in favor of the oil-producing country. As the bargaining power of the host grows, the firm is exposed to an increased degree of political risk. The less vital a company is, the easier it is—and the more likely—for the government to escalate its demands. This process often culminates in either nationalization or divestiture in the face of insurmountable demands by the host state.

To the extent that the firms are able to remain indispensable to operations, they can influence the bargaining process. The concern over the transfer of technology is relevant to this point. Host governments increasingly are demanding access to the technology developed by the Western firms. The companies, on the other hand, have a vested interest in protecting against the transfer: it ensures them an area in which they have control over a vital input, thus affording them a measure of bargaining clout. Over time, of course, technology invariably is disseminated and becomes common industry practice. Hence, it is logical for the companies to push ahead constantly and stay on the frontiers of new breakthroughs.

The pattern of nationalization and state participation is the mirror opposite of the important role played by the foreign firms in the various stages of the industry. Almost without exception, domestic distribution (including pipelines)—the easiest point of entry into the industry—has been the first target of nationalization in the industry. The Iranian and Indonesian national oil companies, for example, took over local marketing in 1955 and 1956, respectively. The next aspect to be taken over has been the smaller, less sophisticated refineries, which usually service the domestic market. Once the operations of these are mastered, complex export-oriented refineries become targets. The last assets nationalized are those used in the production of crude, which traditionally has been the main barrier to entry.

There are, of course, mediating variables other than technology in the transition of bargaining power from firm to host. Many host states lack the infrastructure and indigenous skilled labor force to develop and operate an oil industry. Western managerial skills and, most important, market outlets are another front on which the firms are able to provide needed services, thus ameliorating their decline in influence. As evidenced by the temporary appearance of available excess crude and the slipping of spot prices in late 1981 and early 1982, weak market trends also reduce the bargaining power of the exporting countries.

A logical corollary of the increasing bargaining power at the disposal of host governments, empirically verified by market trends during the past 50 years, is the change in demands on the part of the hosts. Other things being equal, host governments are going to seek to maximize the benefits their countries realize from resource projects. This translates into increasing demands on the oil companies over time and an ongoing effort to take control of the industry.

In general, the most persistent demands on the firms have been made by those countries with the greatest bargaining leverage at the times when political and market conditions accentuate their bargaining power. Nationalization, for example, occurred first in those countries—Russia in 1917 and Mexico in 1938—that had the indigenous capabilities to operate an oil industry, that consumed a significant share of their production internally (that is, were not totally reliant on production for exports via the international companies), and that were not economically dependent on crude exports. Conversely, countries that have minor production or unproved potential are in weaker bargaining positions and cannot command the terms given to major oil producers.

Commodity-producer groups—of which OPEC is the model—are a tool by which the producing countries yield more influence over the industry. The logic is simple: the whole is greater than the sum of its parts. By joining forces and giving mutual support to each member in its dealings with the international companies, the OPEC nations are stronger than their individual positions might warrant. This prevents the firms from using the divide-and-conquer strategy with which British Petroleum frustrated the Iranian nationalization in the 1950s. In fact, this tactic was used effectively against the companies by Libya in 1970, indicating the swing of the pendulum of power in the world industry in the course of less than two decades.

The seemingly inexorable process of change swept away the remains of the traditional concession arrangements in the Middle East and North Africa. The countries achieved control over pricing and production, thus forcing the oil-producing companies to cede their last vestiges of ownership rights. The extent to which the various oil-exporting countries control their oil industries varies, depending on their political goals and bargaining power. None of the oil exporters are completely independent of the Western oil companies, and perhaps none ever will be. Some of the countries recognize the value of their association with the transnational oil companies. Even if host governments are willing to concede a permanent role in their oil industries to the oil companies, the host states will strive to the greatest extent possible to exercise total control over their interests in the world industry.

Resource Nationalism

Nationalist sentiments often have focused on the natural resource wealth of a country, particularly in those states that are minerals exporters. Resources are considered the common patrimony of the land and its people, whether they are seen as a gift from Allah or simply as a vast reservoir of potential wealth. Net outflows mean a continuous depletion of this reservoir, in return for which host governments want appropriate payment. Every barrel or ton exported means that much less left in the ground, that much less for domestic consumption, that much less to pass on to future generations.

Nationalism is a term laden with emotional and psychological undertones, and it is often identified with the irrational. It is more constructive, however, to view nationalism as a legitimate expression of the sentiment, the feelings, and the commitments of a people, of their sense of identity and pride. Nationalism is an understandable phenomenon. Although nationalism seems at times to border on the irrational and to flirt with self-destructive excesses, such instances are the exception.

Resource nationalism comprises the bundle of domestic concerns involved in the development of a country's mineral wealth. In its most tangible form, it is embodied in the concept of permanent sovereignty over natural resources. Permanent sovereignty is the doctrine that each state should enjoy control

> *over its natural resources and all economic activities. In order to safeguard these resources, each State is entitled to exercise effective control over them and their exploitation with means suitable to its own situation, including the right to nationalization or transfer of ownership to its nationals, this right being an expression of the full permanent sovereignty of the State. No State may be subjected to economic, political or any other type of coercion to prevent the free and full exercise of this inalienable right.* [5]

Implicit to permanent sovereignty is the state's right:

> *a) To regulate and exercise authority over foreign investment within its national jurisdiction in accordance with its laws and regulations and in conformity with its national objectives and priorities. No State shall be compelled to grant preferential treat-*

ment to foreign investment; and b) To regulate and supervise the activities of transnational corporations within its national jurisdiction and take measures to ensure that such activities comply with its laws, rules and regulations and conform with its economic and social policies.[6]

The thrust of the movement for permanent sovereignty is largely expressed in the Third World's demand for a new international economic order and the restructuring of the North–South relationship. Although the powers inherent in permanent sovereignty present far-reaching political risks to Western firms, it is readily understandable why they are important to host governments. From the point of view of the oil- or mineral-producing country, permanent sovereignty is a declaration of its political right to exercise total control over its natural resources and is part of the transitional process toward actual control.

The doctrine of permanent sovereignty is more than an abstract assertion of demands. To a large extent, the assertiveness of host government control over its oil has been in accordance with the principles of this doctrine. The founding of OPEC, moreover, has been described as the first instance of "practical significance" in moving toward permanent sovereignty, and the "effective practice of this right was the real impetus for all recent radical developments that have occurred in the international oil industry."[7]

Permanent sovereignty, as the previously referenced United Nations documents indicate, justifies a broad range of government behaviors on the grounds of asserting national control over natural resources. Collective action; control of pricing and production; expropriation and participation; renegotiations; and administrative regulations are but a few examples of the policy changes that can be justified under claims of permanent sovereignty. All these changes present political risks to the oil firms.

In most instances, resource nationalism has been channeled toward satisfying government policy goals. Considering the interests of the oil-producing countries, this has been a rational process. With a handful of exceptions, the governments have pursued the goal of an orderly assumption of control over the oil industry. The phasing in of participation rather than the abruptness of sudden nationalization, the granting of service agreements, and the cultivation of national oil companies are indicative of a paced and planned movement toward permanent sovereignty. Although the oil companies, on the whole, have been forced to beat a hasty retreat, this has been an understandable—and therefore predictable and manageable—process.

There is a less savory side to resource nationalism. In addition to the positive stimuli stemming from the desire to gain control over natural resources, there have been incidents of nationalism as a reaction against foreign companies.[8] Foreign corporations have often been spotlighted as targets of hate against which a government can engender domestic support. As highly visible entities, foreign firms are convenient scapegoats. The antiforeign sentiments that have been aroused on occasion sometimes have swept away investments abruptly, severing mutually constructive ties between the host state and the foreign companies.

Other than as a short-term safety valve for letting off domestic pressure, this nonrational sense of nationalism does not serve the policy interests of the government. As such, it is not the normal channel for the expression of nationalist sentiments; but it does represent a political risk that, though less likely to occur than rational expressions of resource nationalism, is of considerable consequence.

Developed and Developing States

There is an understandable tendency to identify political risk with the developing countries. Third World governments are seen as less stable and more ideologically motivated. Developing countries are more likely to be subject to sudden economic, social, and political changes that may translate into corporate political risks. Nationalism, moreover, often is considered a Third World phenomenon with little application to the West.

In many respects, this distinction between developed and developing countries appears unfounded. Political risks are often present in developed market economy countries in a manner similar to that in developing states. Resource nationalism has emerged as an important political force, in Australia and Canada, for example—the two leading net mineral exporters in the West. Both countries have taken steps to limit the operations of foreign natural resource firms and to assert greater national control over resource development.

"If Canada can be classed as a country where there is a high political risk," *The Times* of London lamented, "it does not leave much hope for the rest of the world."[9] Yet the recent Canadianization and other policies of the Canadian National Energy Program unveiled in 1980 are ample evidence that forced divestiture and government intervention in the oil industry occurs in member states of the Organization for Economic Cooperation and Development (OECD). Although the Canadian government has not been overly enthusiastic about funding the movement toward the goal of domestication of at least 50 percent of the nation's oil and gas production by 1990, there were almost a dozen acquisitions of foreign-owned oil firms within one year of the launching of the program.[10] (None of the companies acquired were among the ten largest foreign producers, among which production is concentrated.)

Seemingly domestic political squabbles in the developed states can have important repercussions for the oil companies. Following publication of Prime Minister Trudeau's energy plan, there was a sharp backlash by a number of western provinces that were dissatisfied with the proposed prices and division of taxes and control between Ottawa and the provincial capitals. Alberta's Premier Lougheed, for example, threatened to curb local oil and gas production by 15 percent if policy concessions were not made to Alberta's interests.[11] Although the Alberta–Ottawa conflict was partially smoothed over and no production cuts occurred, the incident illustrates the types of potential risks to firms that stem from domestic controversy.

The United Kingdom and Norway, the leading net oil exporters among the OECD members, also have engaged in a number of policy changes with adverse repercussions for the oil-producing companies. Since 1974 the governments in both countries have raised taxes, added new taxes, and altered the investment regime a number of times. The result has been additional costs to be borne by the companies. The firms have not challenged the competence of the British or Norwegian parliaments, however, and the changes are not considered manifestations of political instability. Although the oil companies have protested against the changes, their complaints seem to be different from those directed at Third World governments.

This is not surprising. Besides the common traditions and Western values among the majority of the OECD countries, these states tend to have well-institutionalized political structures and processes. The oil companies understand and are able to work with the political processes in these states. Although parliaments differ from the U.S. Congress, they are legislative bodies with well-defined traditions, rules, and procedures. The oil companies know how to mount publicity campaigns for and against government policies in Western states.

Similar to their corporate activities in the United States, U.S. companies operating in other Western countries are able to lobby decision makers, use the media in their efforts to get their position recognized, and perhaps influence the government. By cooperating in interest groups, the oil firms are better able to aggregate and articulate their interests, thus increasing the likelihood that they will have an impact on political decisions. These activities, all of which are common to the Western democratic traditions of the market economy countries, have been used by American oil companies in the United Kingdom, for example. To register their dissatisfaction with the Supplementary Petroleum Duty (SPD) imposed by the British government in March 1981, for instance, U.S. oil firms worked with the two leading British oil industry interest groups, the U.K. Offshore Operators Association (UKOOA) and the Association of British Independent Oil companies (Brindex), in their efforts to change the national oil regime in the North Sea.

Denmark, a much less significant North Sea producer, has followed in the steps of the United Kingdom and Norway. Taxes have been raised, and the Dansk Undergrunds Consortium (owned by AP Moller, Shell, Chevron, and Texaco and operated by AP Moller) was compelled in 1981 to cede its exclusive offshore rights to avoid possible expropriation.[12]

Even in the United States, the tide of political events is of vital importance to the performance of the oil companies. Windfall profit taxes, cuts in depletion allowances, divestiture proposals, and price controls on crude oil and natural gas are obvious examples of issues whose resolution affects the interests and profits of the petroleum firms. Less well publicized, however, are potential problems stemming from the efforts by a number of states, including California, Connecticut, Vermont, and Wisconsin, to tax the corporate profits earned from oil operations worldwide.[13] The outcome of court suits may have adverse economic effects on the companies' short-term economic interests and may generate a political climate that threatens their commercial interests both in other states in the United States and abroad.

As we have indicated, risks are not limited to a stereotypical "banana republic" of any sort. The oil industry is particularly vulnerable to political actions because of its relative and absolute importance in the economies of even the economically diversified Western countries. Risks are evident in both the developed and the developing nations and are not necessarily greater in the latter than in the former. Corporate perceptions and abilities to manage the risks, however, do differ between the classifications of countries.

In contrast to the ability of the oil companies to participate in the political processes of the Western democracies, Third World countries usually have less well developed political structures and channels for foreign companies to influence political decisions. The often restricted nature and circulation of the press and other media restrict publicity efforts and similar problems confront lobbying plans. Many developed countries, moreover, are not elected democracies. These highly personalized regimes lack domestic legislative and democratic traditions. Coping with government by fiat presents many difficulties to oil companies and others. Both currying and failing to curry the political favor of such regimes carry risks.

The Political Economy of Oil and Gas Agreements

Transnational companies have displayed a natural tendency to generalize from Western legal and political traditions and institutions expecting parallels in other parts of the world. In dealing with foreign governments, however, many companies have learned that the rules of the game are different. Agreements between a sovereign state and a private firm are significantly different from agreements between two private entities. Moreover, the domestic regulations and norms governing contractual relationships have limited applications internationally, particularly when one party is a foreign government or its representative.

Governments have the authority to violate or terminate agreements—and they can be expected to do so. As a rule, however, most agreements will be heeded, if for no other reason than that it is in the government's interest or that it might avoid a loss of credibility in negotiating future agreements. The nature of sovereignty and the state are such, however, that states can be bound neither by simple contractual arrangements nor by limited private interests. Even in the West the state has the recognized right to expropriate property or otherwise violate property rights (including contractual rights) in the exercise of police power (in which event no compensation is required[14]) or in the name of the national interest (which requires compensation), however it is defined.

Unaccustomed or unwilling to think in terms of politics, executives of many firms see government termination, violation, or alteration of agreements as a juridically defined breach of contract. The sanctity of property and contract, which may not be the norm in another culture, is not sufficient to override the interests of state. It is understandable why foreign companies believe that host governments should provide for continuous and stable investment regimes, should not discriminate against foreign firms, and should abide by the terms of an agreement. It is politically naive, however, to think that governments always will behave thus.

Part of the drive for permanent sovereignty over natural resources and the desire for a new international economic order has been an assault upon traditional international laws and economic norms. Mineral- and oil-exporting countries consider that the international economic system is structured to their disadvantage. Norms taken for granted in the United States, such as private property and sanctity of contract, often are seen as prejudicial to the interest of other states. Supporting such rules, the argument runs, is not neutral, as the private companies would have us believe; rather, it is reinforcing the status quo, which many of the oil- and resource-exporting states see as unacceptable and unfair.

The shifting of bargaining powers between host and firm, to which we alluded previously, has prompted what most firms have considered to be grossly unjust violations of agreements. Host governments, on the other hand, see the original agreements as unfair and consider that changes in midstream are the appropriate resolution.

Time after time, the political economy of what has been termed the "concession process"[15] has brought about the assertion of state prerogatives in unilaterally ending or alternating agreements. The concession arrangements that were swept away in the early 1970s, although none of them had expired, had outlived their usefulness to the host governments and no longer reflected the balance of bargaining strengths held by host and firm. The majority of the Venezuelan concessions were binding through 1981; the Aramco agreement was protected until 1999; and the Kuwait Oil Company (KOC)—owned 50 percent each by Gulf and BP—was not to expire until 2026. Yet the Venezuelan concessions were fully nationalized in 1974–1975; Saudi participation in Aramco commenced in 1973 and reached 100 percent in 1980; and KOC was 100 percent nationalized in 1975.

In strictly economic terms, "the objective of mineral economies," as described in a World Bank Staff Working Paper, "is simple: it is to capture all rents* while letting the investor make the return necessary to induce him to invest."[16] If a govern-

---

*Rent encompasses all the revenues generated by a project above those needed to induce the initial investment. As such, it is considered surplus in economic jargon.

ment is capable of producing its oil at a rate of profit to the state that is equal to or greater than that earned by permitting foreign investment, it has little inducement to allow foreign equity participation in the national oil industry. This situation should prevail unless the government can earn higher returns by channeling its investments elsewhere or if particular policy concerns recommend partial reliance on the transnational companies.

Interests of Host and Firm

Strict economic terms are never the criteria for decision making, however. Although economic utility functions often are the most important variable in decision making, the decision-making process is political. Moreover, ranking priorities and goals to maximize ends entails attaching political values or judgments and is not an abstract econometric process.

Understanding the political economy between host governments and oil companies requires more than a numerical comparison of the profit maximization function of the two parties. Drawing on our discussion of relative bargaining power, the dynamics of the relationship might best be understood in the context of the respective interests of the government and the company.[17]

Table 2.1 ranks the interest priorities of host governments in comparison with those of the oil-producing companies. The first priority of the government must be to stay in office; without this, all further goals are irrelevant. More important, the sensitive political and development concerns surrounding oil policies often prove important to a regime's chances of achieving popularity, support, or even survival.

Once a regime is secure in power, it can address the economic maximization function. Assuming that the regime is not corrupt, it will think in terms of national benefits (a social utility function), not private gain. Although there tends to be a great

Table 2.1   Ranking of Interest Priorities

Host Government	Oil/Mineral-Producing Firm
1. Stay in office	1. Maximize economic returns
2. Maximize economic returns	2. Insure transfer of funds and supply
3. Further domestic policy goals	3. Retain control over operations
4. Exercise control over resources	4. Retain ownership
5. Further foreign policy goals	

SOURCE: Howard L. Lax, "Natural Resources in International Politics: Conflict and Cooperation," paper presented at the CUNY/Political Science Conference, The Graduate School and University Center of the City University of New York, 11–12 December 1981.

deal of private wealth in the hands of leaders in oil-producing states, we can assume that the focus is more on national or social economic returns than on personal ones.

After increasing the economic returns flowing to the state, the government turns its attention to having oil development serve domestic policy goals. At this stage, the focus on simply increasing the amount of earnings is redirected toward an array of development and other policy goals.

The government's fourth priority entails increased assertion of its control over its hydrocarbon resources, a procedure begun with the preceding priority. The turn toward fulfilling the principles of permanent sovereignty over natural resources is largely embodied in this priority. As a last concern, oil-producing countries can be expected to use their oil wealth for foreign policy goals.

The oil firm's priorities, not unexpectedly, are different. The bottom line is profits, and the fundamental goal of the company is to make money. A related interest priority is the ability to transfer profits abroad in hard currency. The transferability of supply—that is, the ability to export crude—is equally important.

The third preference of the oil firms has been to retain control over operations. This fosters profit maximization and permits rational utilization of the various exploration, production, refining, transportation, and distribution capabilities of the firm. The last priority is ownership. Traditionally, the oil companies retained ownership rights under the concession agreements. This has been a less important issue than many analysts previously predicted, as ownership often proves to have little to do with the bundle of other—and more important—interests of the companies.

When the concessions were initially granted, mostly in the interwar period, the oil companies were able to satisfy all their interests, while the local leaders solidified their power base (priority one). As colonies or newly independent countries, many local regimes relied on the support from foreign governments and the vast new wealth received as payment from the petroleum firms in exchange for concession rights and as payment of royalties. In this time period, the social welfare function and other concerns of the people were largely ignored as the local leaders grew rich and powerful.

As the producing countries began to mature politically, economically, and socially, the structure of the world petroleum market changed, and the bargaining leverage of the host governments began to grow, the demands of the host states changed. The interests being addressed by the oil-producing governments became increasingly divergent from the goals of the firms. The first major shift was to perceive the economic maximization function in terms of social needs. Although the leaders still continued—and continue—to net tremendous personal wealth from oil production and exports, there emerged a sense of social utility, which needed to be promoted by the development of oil resources.

Profit maximization is not a zero-sum game, in which the gains realized by one party necessarily imply a loss of equal magnitude for the other party, because there may be a "joint maximizing function."[18] Although an expanding pie can mean increasing rewards to all parties to an agreement, conflict about the relative distribution of benefits is a fixed-sum situation: if a host state levies an income tax of 80 percent and a royalty of 20 percent, that leaves the producing firm with 16 percent of the profits (less any additional fees deducted from income).* Although higher crude oil prices mean that company profits are far larger today than they were in the years when taxes and royalties (and prices) were much lower, the firms have been forced to yield an increasing share of the returns to the host countries. The host governments continue to seek increased revenues by raising crude prices or by reducing corporate profits. Both strategies present obvious political risks to the firms.

Pursuit of the other interests of the host government—the furthering of domestic policy, national control over natural resources, and foreign policy goals—often conflicts with, or at best accommodates, the interests of the oil companies. Host government priorities—such as government-established below-market-value prices for domestic oil sale in Ecuador, Libya's production cuts in 1980, or Nigeria's expropriation of BP in 1979 (because of reports that the firm had sold Nigerian oil to the Republic of South Africa)—have increasingly imposed additional costs and political risks on the companies.

This state of affairs may be expected to continue. The priorities ranked in table 2.1, which are based on empirical evidence as well as on normative and rational deduction, are an indication of the extent to which the general goals of oil-producing countries differ from the interest of the oil companies. The more aggressive the producing countries are in pursuing priorities beyond the minimalist desire of staying in office, the more the host interests are likely to diverge from those of the oil companies.

Rational development goals induce countries to relate oil production to the pace of national development and economic growth. Ceilings on Mexican oil ouput, for example, are not expressions of anti-Americanism; rather, they represent a serious effort of the development-minded Mexican government to avoid flooding the Mexican economy with vast amounts of currency that cannot be absorbed productively by the domestic market. Because the social discount rate of the country often differs from the economic discount rate of the firm, the production schedules desired by the parties are unlikely to be similar.

From the host's point of view, leaving the crude in the ground not only may result in higher earnings through the appreciation of prices, it also affords the state a measure of bargaining insurance and provides for the welfare of future generations.

---

*From a hypothetical after-cost income of 100 units, 20 percent (20 units) goes to royalty. As royalties are tax deductible, 80 percent of the remaining 80 units (64 units) is paid as income tax, leaving 16 units (or 16 percent of the original sum) for corporate profits.

By viewing a nation's natural resources as part of the capital stock of the country, we can understand better the importance of resources to future development. Oil is an illiquid capital asset that wants conversion to a liquid form. Once consumed, that volume of oil is not available for future use. If the revenues from sales are immediately consumed rather than invested and accumulated, the country eventually will deplete this capital stock, the principal one in Third World oil-producing countries, and will be left without a source of income. If oil revenues are employed in economic development, the state is investing in its future, when its oil reserves and earnings may begin to decline. This not only reflects on the importance of crude to national development but also sheds some light on nationalist aspirations to avoid foreign ownership and control.

Long-term host government goals are even more incompatible with those of the oil firms. Given their own way, the governments would become totally independent of the oil companies in all aspects of operations and would consume most of the state's production domestically. Although it is highly unlikely that the oil producers ever will achieve 100 percent control over the world oil industry, host state goals will continue to run counter to the interest of the oil companies.

Home Governments and Oil Companies

Less obvious but still important to the oil firms is the degree to which the petroleum priorities of their home governments no longer are congruent with the companies' interests. Historically, the interests of the companies and the U.S. government were assumed to be similar—and usually were. The government has a sense of priorities, however, that is distinct from that of the oil firms. In the United States, government goals with respect to international petroleum (and other resources) issues are ranked in the following order:

(1) further foreign policy goals;
(2) insure access to supply;
(3) promote the interests of U.S.-domiciled firms; and
(4) minimize the cost of supply.[19]

Until the 1960s, U.S. foreign policy goals often were tantamount to supporting the interests of American oil firms. This is no longer the case, however. President Nixon put the maintenance of a good relationship with Peru ahead of corporate (Exxon) interests in receiving ''prompt and adequate compensation'' for the expropriation of the International Petroleum Company in 1968; President Ford imperiled Gulf's Cabinda concession in Angola in 1975–1976 by promoting U.S. involvement in that country's civil war; and President Carter completed the process of Iranian alienation from the United States and recriminations against Western firms by admitting the shah into the country in 1979. The Reagan administration's tentative proposal in December 1981 to boycott Libyan oil is but the most recent example of the political risks to which the oil firms have been subject because of the disparate interests of the companies and the U.S. government. The overall nonaccommodative attitude of the Reagan presidency, toward both the Third World and the communist countries, can be expected to complicate political affairs for the oil companies in other areas of the world over the next few years.

The remaining priorities of the government are less in conflict with and are even supportive of corporate concerns. Even here, however, assured access to supply may run counter to the policies of the companies, particularly in the event of an embargo that the host governments expect the oil firms to execute (as in 1973-1974). In the Western European nations and Japan, moreover, where access to supply takes precedence over foreign policy, assuring oil imports has prompted a much greater government role in the industry and has spawned consumer-country national oil companies. Although a U.S. national oil company seems unlikely, the idea has been raised periodically ever since the Roosevelt administration became concerned with the vital nature of oil supplies in maintaining the war effort.

Notes

1. For the text of the Declaratory Statement, see Dankwart A. Rustow and John F. Mugno, *OPEC: Success and Prospects* (New York: New York University Press for the Council on Foreign Relations, 1976), pp. 166–169.

2. F.T. Haner, "Integrating Political Risk Assessment with Economic Evaluations," in *The Measurement of Political Risk and Foreign Investment Strategy*, (proceedings), ed. Dan Haendel, Gerald T. West, and Robert G. Meadow (Philadelphia: Foreign Policy Research Institute, 1975), p. 64.

3. Howard L. Lax and W.G. Prast, "Foreign Investment Mining Agreements in the Late Seventies," paper prepared for the United Nations Centre on Transnational Corporations, 1981, p. 10.

4. Albert O. Hirschman, "How to Divest in Latin America and Why," in *The Multinational Enterprise in Transition*, ed. A. Kapoor and Phillip D. Grub (Princeton: Darwin Press, 1972), p. 449.

5. United Nations General Assembly Declaration on the Establishment of a New International Economic Order, May 1, 1974, cited in Paul Rogers, ed., *Future Resources and World Development* (New York: Plenum Press, 1976), pp. 135–139.

6. United Nations General Assembly Resolution on the Charter of Economic Rights and Duties of States, December 12, 1974, cited in Zuhayr Mikdashi, *The International Politics of Natural Resources* (Ithaca: Cornell University Press, 1976), pp. 17–18.

7. Fadhil J. Al-Chalabi, *OPEC and the International Oil Industry: A Changing Structure* (Oxford: Oxford University Press for the Organization of Arab Exporting Countries, 1980), pp. 1, 67.

8. Luis Vallenilla, *Oil: the Making of a New Economic Order: Venezuelan Oil and OPEC* (New York: McGraw-Hill, 1975), pp. 53–54.

9. *The Times* (London), 10 February 1981.

10. *New York Times*, 5 November 1981.

11. *New York Times*, 31 October 1981.

12. *Petroleum Economist*, April 1981, p. 173.

13. *New York Times*, 26 January 1981.

14. Adeoye A. Akinsanya, *The Expropriation of Multinational Property in the Third World* (New York: Praeger, 1980), p. 235.

15. David N. Smith and Louis T. Wells, Jr., *Negotiating Third-World Mineral Agreements* (Cambridge, Mass.: Ballinger, 1975), pp. 3–4.

16. Gobind Nankani, *Development Problems of Mineral-Exporting Countries*, World Bank Staff Working Paper No. 354 (Washington, D.C.: World Bank, 1979), p. iii.

17. This discussion is based on Howard L. Lax, "Natural Resources in International Politics: Conflict and Cooperation," paper presented at the CUNY/Political Science Conference, The Graduate School and University Center of the City University of New York, 11–12 December 1981.

18. Raymond Mikesell, *Foreign Investment in Copper Mining* (Baltimore: Johns Hopkins University Press for Resources for the Future, 1975), p. 28.

19. Lax, "Natural Resources," p. 17.

3

Substance and Structure of the Oil and Gas Industry

Nature of Investments

Investments in the petroleum and natural resource sectors cannot but be partially shaped by the physical characteristics of the commodities and the nature of the market. The qualities that are inherent to petroleum and natural gas not only render them useful as energy sources but also are important in determining investments in the exploitation and the flow—both literally and metaphorically—of the international market.

Like other natural resources, crude oil and gas must be discovered or found in the ground; they cannot be invented or manufactured. Hydrocarbon reserves are unequally distributed among the regions and states of the world.* Only areas that have the proper conditions, with a history of the necessary geological activity, will be blessed with petroleum or natural gas. Unequal distribution means that production will be concentrated in certain areas and will not be evenly spread between the approximately 170 countries in the world.

Unless the market is disturbed by noneconomic forces (as it invariably is), investments should be patterned like the distribution of reserves. Since oil and gas resources are not randomly distributed, one would expect exploration activity to be

---

*Reserves are the amounts of proved crude and gas that are economically exploitable given prevailing technology and market conditions. Resources is a broader category, which includes both reserves and other amounts that have been inferred, are hypothetical, or are uneconomical to produce under current technological and market conditions.

focused on areas that have the most promising geological history. This seems logical, as resources can be discovered only where they rest in the earth.

This abstract logic, however, fails to provide an accurate description of the actual pattern of exploration and investment. Areas that are judged politically safe have been subject to far more rigorous and systematic exploration than most of the Third World, a large part of which has been underexplored. Exploration in the United States and Western Europe, on the other hand, is at record levels. Accommodating developing countries are beginning to receive attention as potential areas for exploration, but the OPEC countries, most of which have not been subject to exhaustive survey, are suffering from declining exploration investments.

The logic also fails to take into account the sequential nature of explorations and the limited funds available. A firm can pursue only so many exploratory projects simultaneously. Prospects that are less promising for geological, economic, or political reasons receive less attention and are unable to compete for the firm's limited human, material, and capital resources.

Exploration is very risky and expensive and can account for 20 to 25 percent of the total costs involved in a project. Most host governments are unwilling or unable to assume the high degree of risk associated with exploration and prefer to rely on the transnational oil companies to carry the risks of exploration. In effect, the firms are better able to bear the risks because they can pool the risks from a number of projects. This affords a measure of insurance against the large number of unsuccessful, costly exploration efforts. In addition to their ability to pool risks, the oil companies command the necessary inputs and technology and often enjoy a healthier cash flow than Third World host governments.

Oil and gas production are highly capital intensive, requiring huge capital outlays. Costs vary widely, depending on the location and geological structure of the crude or gas field being developed. Since the easiest-to-develop fields tend to be the first ones explored and brought into production, there is a trend toward increasing costs. This is a function of the less-hospitable and less-accessible locations of new fields, which often are offshore, in the depths of the jungle, or in frigid climes. Preproduction costs—including feasibility studies, exploration, infrastructure, and construction—have escalated rapidly in the past few years and show no sign of lessening.

Actual production costs are highly variable. The extraction of a barrel of crude may cost as little as 50 cents in Saudi Arabia and as much as $15 or $20 in the Arctic. Lower-cost operations therefore yield wildly extravagant rents for the host countries. Although the rents earned by low-cost producers bear no functional relationship to the costs of production, the high-cost projects may operate at the margins and earn minimal returns. Production, moreover, lends itself to large economies of scale.

Infrastructure costs, in particular, have risen dramatically. Pipelines, which are especially important for natural gas projects, have proved to be increasingly costly.

The 1200-mile trans-Mediterranean pipeline from Algeria to Italy was completed in 1981 at an estimated cost of $4.5 billion. The planned 3600-mile Yamal pipeline, which will bring Soviet natural gas to Western Europe, has an estimated price tag of more than $30 billion, while the proposed 4800-mile Alaskan gas line is expected to cost over $40 billion.

Once production commences, virtually all of the investment is sunk cost, which is irretrievably in place. Unlike a factory, a well cannot be dismantled and shipped elsewhere. An operating hydrocarbon-producing project is a permanent hostage in the hands of the host government, which then has not only the fuel in the ground but the means to lift it. Although the geological and economic risks of a project—which the host wants the oil companies to assume—have all but vanished by the time production commences, the polticial risks confronting the firm become intensified.

The huge investments, the long lead times in commencing production and generating income, and the fixed nature of the capital invested make the stability and permanence of the investment regime especially important to the oil companies. Simultaneously, this leaves the firms highly vulnerable to changes imposed by the government. This tends to be reflected in discounted cash flow projections, including political risk discounting by the firm. Extended payback periods are enticing targets for host governments, because during that time the investment is largely completed, the economic and geological risks of the project are sharply reduced, and host profits may be perceived as too small, while corporate returns may be seen as too high.

Hydrocarbons are depletable, and they currently have few adequate substitutes. Because oil and gas are nonrenewable, the pace of production and consumption is an important issue: whether to conserve for tomorrow or to produce and consume today while looking for more reserves or substitutes for the future. Combined with the uneven global distribution of reserves and consumption, the depletable nature of hydrocarbons is the primary source of the sense of scarcity that pervades serious thought about the world industry. The problem is magnified by the unrecoverable nature of disbursed energy, which prevents any secondary recovery or recycling of petroleum and natural gas, as is common with hard minerals.

The exhaustibility of fields inevitably creates concern about the pace and schedule of production, especially for states that have reserves that may approach depletion before the close of the century. Even for countries such as Saudi Arabia and Kuwait—which have known, proven reserves sufficient to last, at present rates of production, until the middle of the twenty-first century or beyond and suspected, unproven resources that will last much longer—there is a sense of finiteness to the supply of oil. Eventually, workable fields become increasingly scarce and the costs of extraction multiply beyond economic alternative energy sources. Investments in secondary and tertiary recovery, which extend the life of a field and may increase the volume of crude extracted by 20 to 25 percent, are efforts at postponing the inevitable. The frontiers of additional recovery, however, are an important area in which the oil firms have an exploitable technology edge that they can use to their advantage in their relationships with host governments.

As a highly flammable liquid, petroleum is very difficult to store and transport. Storage and transportation are even greater problems with natural gas. Unlike most other fuel and nonfuel minerals, which can be stored in bulk piles for extended periods of time with little loss, crude oil and natural gas require special handling and storage to avoid loss and accidents.

Natural gas and crude often are coproduced. This allows sharing of drilling, production, and, to a lesser extent, infrastructure costs. The accounting of the projects permits the assignment of costs in whatever manner is the most profitable. A significant percentage of costs will be fuel-specific, especially if the gas is liquefied for export, but shared costs are common to the industry.

Refinery investments also are partially conditioned by the physical nature of the product. The movement after World War II, away from source refineries—facilities located near the crude production sites—to refineries upstream near the consumption centers reflects the economic logic of shipping the raw commodity rather than finished products.* It is far less costly to ship large volumes of crude than small amounts of products. The increasing capacity of tankers contributed to the logic of shipping crude to refineries near distribution points.

Evolution of Industry Structure

Industry structure is both a reflection and a determinant of the bargaining relationship between host and firm. As the bargaining strengths of host governments have waxed and those of the oil companies have waned, the structure of the industry has responded to distribution of market power. Simultaneously, the changing market structure, on the whole, has promoted the bargaining leverage of the oil-producing countries. Such structural variables as degree of concentration, availability of needed skills and technology, extent of vertical integration of a particular firm, and the foci of control over production affect and are affected by the bargaining process.

The historical structure of the industry, which included a high degree of concentration, operational and discretionary control by the oil companies, extensive vertical integration, and a near monopoly by the companies on the ability to command the inputs necessary to produce crude and run the industry, mirrored and supported the dominant position enjoyed by the companies. The diffuseness of the contemporary structure, which exhibits a proliferation of the number of participants in all stages of

---

*Before World War II, almost 70 percent of the oil traded internationally was refined near the site of production; 40 years later, more than 85 percent was refined in the consuming countries.

the industry, and the control wielded by OPEC and other oil exporters stems from and contributes to the ascendant bargaining strengths of the host governments.

The changes in industry pricing are indicative of the evolution of the industry's structure. Under traditional concession agreements, prices were the prerogative of the oil-producing companies and essentially were irrelevant to the host governments, whose earnings were not pegged to prices or company profits. As royalties were changed from being a function of volume to being a percentage of income from sales, and as income taxes were assessed on company profits, prices became an important determinant of host country earnings. The countries did not, however, have any say in determining price levels.

Throughout the 1950s, the countries expressed their desire to have a voice in setting prices, but to no avail. The cut in posted prices by the major oil companies was the spark that stimulated the formation of OPEC in 1960. The organization continued to insist that the governments have some input in determining prices. Although the companies continued to refuse to cede their control over prices, OPEC was able to stabilize the posted price—the price on which payments of royalties and income were determined. Thereafter, the companies had to absorb the full costs of any discounts from the posted price while continuing to pay royalties and taxes based on a higher price.

The 1968 Declaratory Statement codified OPEC's belief that tax reference prices should be determined by the host government in conjunction with other members of the organization. The producing states' pressure increased in 1970, when Libya began to push the independent operators in the country to accede to price increases. The following year, the host/company meetings in Tehran and Tripoli were the first instance where the companies recognized the right—and the power—of the countries to play a role in negotiating oil prices.* The efforts by the firms to secure a five-year agreement in Tehran "can be seen as a defensive effort to reestablish some measure of stability in a situation in which their control seemed to be ebbing."[1]

The governments maintained constant pressure on the firms to increase prices. Following the devaluation of the dollar in 1971, the OPEC countries had little trouble getting the oil companies to agree to an adjusted price formula, restoring the real (inflation-discounted) price. All the while the companies were making price concessions, the OPEC governments pressed new pricing (as well as other) demands.

The watershed mark with respect to control over prices (and production) was reached in 1973-1974. The Organization of Arab Petroleum Exporting Countries

---

*Moreover, the Tehran and Tripoli meetings were the first time the companies recognized the group negotiating rights of the OPEC countries. Previously, the companies insisted on negotiating with each country individually.

(OAPEC) embargo swiftly turned a buyers' market into a sellers' market. Prices multiplied fourfold as the host governments terminated negotiations with the companies and decided to set prices unilaterally. Prices continue to be the exclusive domain of the governments, with the companies relegated to the role of making yes or no decisions on whether to buy at a given price.

Prices on buy-back, phase-in or bridging, and equity crude were somewhat negotiable, as were prices of crude bought under contract.* Official selling prices are wholly within the discretion of the host government, as are discounts from and premiums in addition to the price. As might be expected, prices rise more easily than they fall. This upward stickiness of prices, better known as the ratchet effect, is maintained by the aggressive pricing strategies employed by the oil-exporting countries. When the market remains in a constant state of oversupply, however, as it did for a large part of 1981 and 1982, prices may come down.

Independent of the control of seller and buyer, the spot market plays an important pricing function with respect to crude oil. As a barometer of supply/demand forces and uncertainty on world markets, the spot trade, largely concentrated on the Rotterdam market, is a sensitive indicator of the prices buyers are willing to pay and sellers are willing to accept. In the last few years, the spot market has handled increasing volumes of crude being sold by the OPEC member governments to test what the market will bear. Although spot prices sometimes prove to be hypersensitive in fluctuation, they also have been good indicators of world price movements.

Following the Iranian revolution, the spot trade became more brisk as prices climbed. Kuwait, for example, was reportedly selling as much as 700,000 barrels per day—approximately 45 percent of output—on the spot market in 1980.[2] Contract and official prices soon followed the trend set on the Rotterdam market. Following the assassination of Egyptian president Anwar el-Sadat in the fall of 1981, however, spot prices rose only moderately and for only a brief period before falling back in line with prevailing official selling prices.

The softening in world oil prices in the second half of 1981 and early 1982 was reflected in the spot trade. While the OPEC countries struggled to keep prices from falling by setting physical production levels for members (for the first time), spot prices slid dramatically in response to the oversupply of crude in the international market. By the end of March 1982, Rotterdam prices for Saudi light were down to $26 per barrel—$8 below the official marker price.

---

*Host government participation in the industry necessitated the creation of various price categories for crude sold to the companies. Buy-back crude—the volume of production owned by the government (by virtue of participation) to be marketed by the companies—was priced at 93 percent of the posted price. Phase-in or bridging oil, which was to help the firms fulfill their contractual obligations to provide supply was sold at a somewhat lower price. Equity crude—the amount of output to which the companies were entitled because of their share of ownership—was priced at approximately 80 percent of the posted price, depending on the rate of royalties and income taxes.

In contrast to OPEC pricing policy, the British National Oil Company (BNOC), which sets British North Sea prices and is the price leader for North Sea producers in general, has proved more sensitive to spot price trends. As Rotterdam prices fell, BNOC cut the price of its high-quality North Sea crude to $31 per barrel, significantly undercutting the inflexible Nigerian price of $35.50 per barrel for similar quality crude. Conversely, as the market firmed, by the close of May 1982 spot prices for North Sea oil had risen $3 to $4 above the BNOC price. In response, BNOC and the other North Sea producers raised prices by $2.50 per barrel in June 1982.

Natural gas and liquefied natural gas (LNG) prices in world trade have a shorter history than petroleum prices. There is no spot market for natural gas or LNG, and prices are determined by direct negotiations between buyer and seller. Crude prices historically determine the general trend in prices for these commodities. Natural gas and LNG prices vary widely between buyers and sellers, and the trade in both is very young.

The decline in the companies' control is evidenced by the decreasing share of world trade and production they handle. In the early 1970s, the international firms accounted for approximately 90 percent of the noncommunist trade in crude worldwide; by the close of the decade, the share handled by the companies had slipped to less than 50 percent. Most of the firms have lost their concessions and no longer have significant volumes of equity crude, and they are increasingly reliant on buy-back oil. Although it is more costly than equity oil, buy-back crude still gives the companies a price advantage over those buying on the spot market or at official selling prices.

As the volumes handled by the oil companies have declined, they have made fewer sales to third parties. While the majors sold between 6 million and 7 million barrels per day to third-party buyers in 1970, the amount dropped to 3.7 million barrels per day in 1978 and to less than 1 million barrels per day in 1980.[3] BP, for example, once was crude-long (that is, had equity crude in excess of its needs) and sold large volumes to third parties. Following the firm's losses in Iran, Kuwait, and Nigeria, however, BP is crude-short and has trouble securing sufficient supplies to satisfy its needs. As a group, the international firms are net buyers of crude. Although joint ventures and service contracts provide money-making opportunities, they also entail reduced access to supply. The volumes offered as buy-back crude have declined steadily, decreasing the amount of crude handled by the companies.

The reduced role of the companies in world trade has been a function of the increased importance of the host government. Overcoming what proved to be perhaps the main barrier to entry in the world industry during the 1960s and the first half of the 1970s, the governments have assumed an increasingly prominent place as sellers of crude. Market outlets no longer are the nearly exclusive domain of the oil companies. Unlike hard minerals, crude has proved to be readily marketable. The relative ease of marketing crude contributed to the early success of production-sharing agreements, under which the state was entitled to a share of output to dispose of as it saw fit (although clauses often were invoked by which the foreign company marketed a significant share of the government's oil). Crude has become increasingly easy to market over the years since the earliest production-sharing agreements. This has contributed to the success host governments have enjoyed in direct marketing of production.

The main tool of government policy in marketing, as well as in other areas of petroleum and natural gas policies, has been the national oil company (noc). Since June 1979, when Gabon organized Petrogab, a state-owned petroleum firm, all of the OPEC countries have established nocs. The 13 OPEC nocs directly market more than 50 percent of OPEC exports, compared to virtually nothing in 1970 and barely 5 percent in 1973.[4] As late as 1975, for example, the United Arab Emirates was able to market only 5 percent of its 60 percent share of production.[5] The remainder was sold back to the companies for marketing. In 1980, on the other hand, the Iraq National Oil Company (INOC) marketed all of the country's output, while CVP (Corporacien Venezolana del Petroleo) handled the majority of Venezuela's exports.

The national oil companies, which date back to the 1950s among the OPEC members and even farther back in some of the Latin American countries and the Soviet Union, are designed as instruments to implement government policies. Originally, their domain was limited to those parts of the country that were not included in a foreign-held concession or relinquished by the concessionaires. Subsequently, the role of the nocs has been expanded to cover all of the host states' petroleum and natural gas reserves and policies in the petroleum and related sectors.

The nocs serve a number of functions. On the immediate and practical level, they are instrumental in supplying the domestic market, providing an educated cadre of skilled nationals that understands the industry from the inside and can effectively handle negotiations with the international firms, and they are the primary medium through which the government executes its petroleum and gas policies and controls the domestic industry. On a more general plane, the nocs are a source of national pride, the means by which governments assert their will to permanent sovereignty, and a standard against which the performance of foreign firms can be measured.

Through their nocs, the host governments have gained immediate access to markets. The availability of buy-back oil—and its price—is a constant issue of concern to the international oil companies. There are more than 90 nocs worldwide, all of which, by definition, are closely affiliated with governments and politics. Increasingly, consumer states have created nocs that bypass the traditional channel of supply from the private firms. The oil companies are playing the dual roles of producers and buyers of crude. Between the producer nocs on one hand and the consumer nocs on the other, however, there has developed a government-to-government trade relationship that totally excludes the transnational petroleum firms. Direct intergovernmental sales reached almost 6 million barrels per day in 1979, and accounted for 60 percent of Iraqi, 40 percent of Iranian, and 30 percent of Libyan exports.[6] The government-to-government market and the reliance on nocs can be expected to grow throughout the 1980s, further eroding the role played by the private firms.

Not only have the producer nocs sought to market increasing volumes of their crude directly, they also are selling a growing percentage of refined products. OPEC members exported 1.5 million barrels per day of products in 1980, [7] and the volume is expected to almost quadruple over the course of the decade. The exporting of products is a logical offshoot of the principles of permanent sovereignty and the 1968 Declaratory Statement, and it is a means of increasing the value added netted by the oil-producing states. The direct exportation of products, moreover, affords the producer countries greater control over the distribution of their oil, giving them more opportunity to favor or exclude specific markets.

Industry Structure, Policy Decisions, and Political Risk

The decisions made by producer and consumer governments and the responses of the firms to these decisions continue to influence the structure of the industry. At the most basic level, the world industry has diffused into a loose array of numerous heterogeneous participants. The importance of the transnational oil firms has declined dramatically as the host government nocs have asserted their claim to market power. The firms are not about to dissolve; to the contrary, they will continue to perform essential functions at all stages of the industry and in other energy-related areas. Increasingly, however, company activities probably will be circumscribed by government policies, reducing the range of opportunities and prerogatives available to the firms.

Ownership rights are almost universally recognized as residing in the host state. As a matter of policy, a number of countries prohibit the participation of foreign capital and the international companies in the petroleum and natural gas industry. Countries that allow equity ownership often permit only a minority foreign stake or are moving toward reduced foreign ownership. Having ousted foreign oil firms or gained national control over the petroleum industry, the governments have imposed strict terms on the companies. Requirements of regular liftings regardless of needs and linkage of company share of production to the amount taken by the nocs are not uncommon. Production cuts also have been imposed at the will of the governments. Purchases under contract are subject to volatility in price and volumes, with renewal at the prerogative of the host government.

The pace of exploration in non-OPEC countries and particularly in consuming states is indicative of company concern with continued availability of supply on acceptable terms. The current glut has partially alleviated the concern about supply in the short term, but the oil companies harbor no delusions about the current oversupply situation being more than a transitory phenomenon. The market is expected to tighten again before the end of 1982, once again renewing emphasis on questions of access to supply. The statement by Petromin's head Abdulhady Hassan Taher in October 1979 that, as market pressures intensify, "new inertia may be developed to determine the allocation of scarce oil, where political and other considerations may override commercial judgements"[8] is merely confirming evidence of the precarious supply situation of the oil companies. The market has slackened since 1979 (although prices still have risen), but renewed supply/demand pressures—which are inevitable, even if only for short periods of time or because of policy preferences of oil exporters—may mean greater reliance on overtly political criteria in decisions about allocating supplies.

The long-term goal of the OPEC states in the area of marketing is clear: they hope eventually to be able to market 100 percent of their exports through the nocs, thus obliterating the role of the transnational oil companies in marketing. The demotion of foreign firms to service-operating companies is far from realized. In fact, the oil companies still retain equity holdings in a number of OPEC countries, including the four African members, and are important vehicles for marketing in most of the OPEC states. The companies still have the best-developed and most-integrated global marketing networks and are uniquely able to smooth short-term supply imbalances in country or regional markets. Nonetheless, there is a clear universal preference by oil exporters to market production through the nocs rather than through the international companies.

Pricing decisions seem to be the product of a complex relationship between market forces, government policies, and intra-OPEC negotiations. Saudi Arabia continues to be the primary actor influencing price movements worldwide, but it has been unable to maintain a cohesive OPEC pricing strategy. By manipulating supply, Saudi Arabia has orchestrated market forces, thus compelling some coherence in pricing, as evidenced by the narrowing of the broad range of OPEC member prices in late 1981. The current pricing structure, however, which set Saudi marker crude at $34 per barrel and allows for a range from $32 per barrel for heavier crudes to $38 per barrel for lighter, low-sulfur, or "sweet" crudes, will last only so long as the soft market keeps a lid on the price hawks.

Pricing issues present a number of sensitive concerns to the companies. As crude prices rise and the willingness of consumer countries to accept price pass-throughs declines, the firms have been subject to a cost/price squeeze that has reduced per-barrel profits (although profits may have grown in other areas or from the selling of larger volumes). Higher prices have translated into greater country earnings. These earnings have been employed by the host governments to gain additional control over and ownership of the national oil industries, and they continue to be a source of market power wielded by the hosts in negotiations with the oil companies.

More important, so long as the price elasticity of demand is less than one,* the exporting countries have the unique opportunity to earn greater profits by selling less. This provides a powerful incentive to host countries to raise prices. Smaller volumes may yield increased earnings for the governments but might mean reduced profits for the firms. More important, smaller volumes will be easier for the nocs to market and will mean less supply available to the companies.

The investment priorities of the companies mirror the efforts to decrease their dependence on the OPEC countries. The companies have begun to focus on those regions and activities (exploration, petrochemicals) in which they feel most comfortable with their competitive advantages. A number of firms have withdrawn from unprofitable markets, although Exxon's recent decision to divest its interests in Libya is the first example of a company voluntarily withdrawing from crude-producing operations.

The degree of vertical integration seems to be ebbing. There is a trend toward the devolution and decentralization of authority and responsibilities. Dirk de Bruyne, president of Royal Dutch Petroleum Company, has suggested that the globaly integrated structure, though logistically advantageous, may be outdated.[9] Confronted with corporate profit squeezes at both the upstream and downstream ends of operations, he has suggested that each project should be economically profitable and justifiable without reference to the global network. Nationalizations have forced the companies to accept that the integrated chain of operations has been successfully challenged by the producer governments. Government inroads in the ownership and operation of refineries and tanker fleets are but further evidence of punctures in global integration. The withdrawal from unprofitable markets—BP and Shell from Italy or Exxon from India and the Philippines, for example—seems to confirm the new emphasis away from the economic logic of integration and focusing on establishing the economic viability of each operation.

The negotiating strength of the oil-exporting countries is apparent in their direct influence on corporate investments. Libya and Algeria have promoted investments in exploration by linking prices to exploration expenditures. To foster investments in Saudi petrochemical and refinery ventures, Riyadh has offered "incentive crude": additional volumes that the companies can purchase, depending on the amount invested. Investments in projects in Jubail and Yanbu by Exxon, Mobil, Shell, and Ashland, for example, are earning the companies the opportunity to purchase up to

---

*A price elasticity of demand less than one means for every unit of increase in price, there is a less than one unit decline in demand. If price elasticity exceeded one, demand would decrease at a rate greater than any price increases. A perfectly balanced market, in theory, would display a price elasticity of one, such that demand would be an exact gauge of changes in price.

500,000 barrels per day (500 barrels for each $1 million invested, on an estimated total investment of $1 billion) in addition to regular supply. Mobil spokesmen have suggested that their investment will mean an extra 1.4 billion barrels of crude that the company can buy over a 15-year period.[10]

The use of positive incentives and negative disincentives by host governments seeking to promote foreign investments probably will spread. The governments have an array of "carrots and sticks" they can use on the companies. It is hoped that the emphasis will be on rewarding investments, as the Saudis have done, rather than on penalizing those that do not invest. In both instances the host governments are flexing their muscles; both the offer of additional crude, even at government-established prices, allows the companies the chance to both earn a return on the capital invested and obtain an additional volume of supply.

Two other important aspects of the world structure that have undergone a considerable measure of change as a result of OPEC policies are refining and tanker shipments. It is more economic to transport large volumes of crude to be refined near the points of distribution than to refine at the source and ship smaller volumes of products to centers of consumption. As mentioned earlier, the host states prefer to export finished products, thus increasing their share of the value added captured by the domestic economy. The building of refineries and related infrastructure, moreover, is part of the overall development and industrialization process sought by the developing countries.

Despite the spate of refinery projects in the Middle East and North Africa in recent years, more than 90 percent of OPEC's oil exports are in the form of crude. Refinery construction continues strong as host governments and the oil companies, eager to maintain crude supply, invest in new capacity in the oil-exporting countries. The 1.5 million barrels per day of products exported by the OPEC nations in 1980 is expected to climb to 3.5 million barrels per day by 1985 and to reach 5.7 million barrels per day by 1990.[11] Over time, this means a further increase in the negotiating power of the oil-producing countries. Similarly, the swelling of government-controlled refinery capacity will reduce the amount of crude available to the international companies and will promote direct government-to-government sales, bypassing the companies.

The world tanker fleet is undergoing a change in composition similar to that in refining. Despite overcapacity in both, government-owned and -controlled tanker capacity also is growing. By having supply shipped in nationally owned tankers, the country can capture the profits from tanker operations and exercise one more degree of control over the industry, including the destination of exports. Kuwaiti contracts, for example, provide that the government can demand that 50 percent of exports be shipped in Kuwaiti tankers, as well as allowing the Kuwait Oil Company (KOC) and the Kuwait National Petroleum Company (KNPC) to refine in the purchasers' refineries, for their own account, a quantity up to 25 percent of the contracted volume.[12]

Contemporary Petroleum Agreements

Two classes of agreements are used in the petroleum and natural gas industries: concessions and contractual arrangements. Concession-style agreements, in which the company has an equity interest in the project and is subject to royalties assessed on the value of production and income taxes on net earnings, is the most frequently used arrangement, being employed in approximately 120 countries.[13] Contemporary concessions invariably include state participation (joint ventures), usually as the majority partner, in contrast to the traditional 100 percent foreign-owned and -operated concession. In most instances, concessions are not the exclusive mode of agreement; rather, countries tend to use both concession and contractual arrangements. British and Norwegian projects in the North Sea, however, all are concessions, with state participation.

Contractual agreements include production sharing, service contracts, and risk contracts, each of which also can be undertaken as joint ventures with the state. Under these arrangements, ownership rights are the exclusive domain of the host state. The archetypical production-sharing agreements are those negotiated by Indonesia, in which costs of development (initially borne by the foreign participants) are recovered from a portion of production. The remaining production is shared or divided by some agreed formula, and taxes are levied on corporate profits. With modifications, production-sharing agreements are employed in such countries as Libya, Egypt, Syria, Peru, Guatemala, Malaysia, and the Philippines.[14]

Risk contracts were pioneered by Brazil and subsequently have been used by Peru. The basic principles of the risk contract are that the contractor assumes the full burden of risk in exploration and development. If a commercial find is made and brought onstream, the contractor is reimbursed for all costs incurred plus interest and/or fees. Payment usually is made in currency, but also can involve crude.

Service agreements entail the performance of particular services—managerial, construction, or exploration, for example—in return for which the foreign firm receives a stipulated fee. The Arabian-American Oil Company (Aramco)/Saudi relationship has moved through the progressive stages from 100 percent Aramco ownership, to concessionary participation, to complete ownership by the state and operation by the firm under a service contract.

Agreements with the OPEC countries appear to be moving continuously closer to service contracts, although many governments are reluctant to assume the associated risks. Imports of petroleum sector services by OPEC countries have grown rapidly since 1973, largely because of the increased reliance on service contracts. Service agreements are not usually classified as foreign direct investment, as they do not involve the traditional aspects of risk and uncertain returns that are fundamental to the nature of an investment. As a direct sale of services, the firm's earnings are not related to profits or return on capital.

A last type of transaction is the straight purchase of petroleum from the host government or its noc by a foreign oil company. This mode of contract is unrelated to the concept of foreign investment but is an important component of the current industry structure. The nonequity risks to which firms with foreign investments are exposed also affect companies buying under contract. Moreover, most of the transnational oil companies purchasing crude under contract once had an equity investment in the countries from which they buy crude. Contractual sales are usually short term and highly subject to changes imposed by the producing countries with respect to prices and quantities.

The array of agreements in the oil industry is far more complex than previously was the case. There is a seemingly endless variety of hybrids, and the same government often uses different modes of agreement simultaneously; similarly, the oil companies have proved willing to participate in a wide variety of arrangements. The style of agreement adopted in each instance is determined by policy decisions of the host state, with the particulars negotiated with the companies involved.

As a rule, the transnational oil firms have been largley excluded from operations in the major oil producers in the Middle East, with the important exception of Saudi Arabia. Among other OPEC countries, there is a mixture of approaches toward the involvement of the TNCs in the petroleum industry. Iran, Iraq, Kuwait, and Venezuela prohibit direct foreign participation in the industry, as does Mexico, the leading non-OPEC oil producer in the Third World. The United Arab Emirates (UAE), Nigeria, Algeria, and Libya permit foreign equity participation in hydrocarbon development projects, while Indonesia's production-sharing agreements reserve an essential role for foreign oil companies.

In contrast to the traditional concessions, the obligations and responsibilities of the firms under contemporary agreements are far more burdensome, and their range of control and prerogative is far more circumscribed. The modern concessions stipulate shorter periods for the exploration and exploitation stages of a project, often specifying the minimum amount of money to be spent by the foreign firm; relinquishment provisions entail accelerated withdrawal, often limiting the concessionaire to 25 percent or less of the original area within five to seven years; royalties and income taxes are more onerous, being as high as 20 percent and 85 percent, respectively, in the OPEC countries; and ownership and control usually are vested in the state.

The political risks confronting companies are different for the various types of agreements. Purchases of crude are subject to what we earlier termed contractual risks. Basically, this covers issues of price variability and access to supplies. Renegotiations of Kuwaiti sales agreements with Shell, Gulf, and BP that expired on April 1, 1981, for example, were delayed because the government insisted on a $5.50 per barrel premium, which the companies were unprepared to pay because of the surplus crude available on the market at that time. Although a price settlement was effected, the three firms were compelled to settle for reduced volumes. Similarly, Iranian sales to Japan were suspended for the majority of 1980 because of a price dispute, as were Algerian LNG sales to El Paso (U.S.) and Gaz de France.

Service agreements are not likely targets of risk. The arrangements involve no capital investments that can be held captive by the host. Based on a *quid pro quo*, they are essentially a sale of services on an ongoing basis. The only potential risks have to do with the transfer abroad of personnel and payment. The host has little motive to block the transfer of either, as such an action would mean suspension of the services provided by the foreign firm.

Under risk contracts, the company must cope with transfer and administrative/statutory risks. Having proved a viable field and brought it into production, the firm has made a sizable investment, which it wants returned plus profit. In addition to potential changes in the host's willingness to transfer abroad such amounts in hard currency, administrative/statutory changes might affect the profitability of a project.

Production-sharing agreements entail all the risks implicit to a risk agreement plus the added dimension of operational risks. Unlike risk or service contracts, under production sharing the companies involved have a vested interest in the day-to-day performance of a project. The ability to recoup the initial investment and profitability will be determined by the administrative/statutory regime and the level of output. The production-sharing partner also may be prey to ownership risks, as the assets may be expropriated or nationalized.

Concession agreements expose the transnational oil firm to ownership risk on a more substantive scale than in production sharing. In both instances, the company makes large investments, which can be seized. Under the terms of a concession, moreover, the company's equity stake in a project is vulnerable to expropriation or nationalization. Concessionaries also must cope with the full range of political risks involving transfer, administrative/statutory, and operational issues.

This is not meant to imply that a concession or a production-sharing agreement is inherently riskier than a risk contract or a service agreement. Rather, the former type of arrangement involves a greater diversity of activities and responsibilities on the part of the foreign company, thereby increasing the potential types of risk involved. Simultaneously, however, the concession and production-sharing arrangements provide more opportunities for the companies involved to exercise a real influence over important decisions with respect to all stages of operations (particularly in concessions) and to net higher rates of returns.

Continuing Role of OPEC

OPEC was born of limited aspirations. The expectations of the original members—Venezuela, Iran, Kuwait, Saudi Arabia, and Iraq—were modest. The transnational oil firms and Western governments were barely interested in the new organization.

Building on a series of limited successes, bringing additional members into its ranks, and steadily increasing its negotiating power and assertiveness, however,

OPEC became the single most dominant force in the world petroleum industry. Its successes, often won at the expense of the oil companies and their home governments, have humbled the former colonial overseers of the member countries and have partially restructured the world economy, creating the largest wave of capital flowing from the developed to the developing countries in history. The organization has been the pride of the Third World (despite the inability of non-oil-producing developing countries to afford their inflated oil bills). Other groups have sought to emulate OPEC's success, the political and economic demands embodied in the call for a new international economic order have been partially stimulated by the organization's experiences, and OPEC members have taken a leading role in the nonaligned movement among Third World nations.*

Whether or not OPEC is a cartel is a question that continues to be mooted. There can be little disagreement, however, that the organization has been instrumental in shifting the scales of bargaining power in favor of the member states. In the process, it has gained control over pricing, production, and most other important aspects of the industry.

A combination of characteristics and circumstances is the source of OPEC's market power. The 13 member countries accounted for 40 percent of world crude production and 54 percent of noncommunist production during the first half of 1981. More important to its grasp on the world market, OPEC countries account for approximately 80 percent of the crude oil that enters into noncommunist trade. The same countries are sitting atop two-thirds of the world's proved petroleum reserves and three-quarters of the reserves in noncommunist countries.

The magnitude of the organization's dominance over the export market is punctuated by the low real marginal costs of production—under $1 per barrel on the average—and the capital surpluses accumulated by a number of oil producers. Some of the local economies lack the capacity to absorb the tremendous influx of wealth. The "saver countries," as they have been termed—those whose populations and economies are too small to consume more than a fraction of their annual oil revenues, such as Saudi Arabia, Kuwait, Qatar, the UAE, and Libya—could afford to cut production without suffering adverse economic affects. Even for countries with less surplus production, oil in the ground often appears to be a better investment than accumulating or consuming excess revenues. This adds a large measure of flexibility to OPEC in the area of production policies.

---

*The nonaligned movement, which dates back to the mid-1950s, reflects the desires of many developing countries to avoid permanent alliance with either the Western or Eastern blocs or camps and to forge an independent world political force representing the interests of the Third World. The Group of 77, which boasts more than 100 members, is the main forum expressing the viewpoints of nonalignment.

The price-inelastic nature of demand has enabled producer countries to realize extravagantly increased revenues per barrel without loss of export markets. The drop in Western oil consumption in 1980 and 1981, which has been estimated at approximately 12 percent for the two years combined, may indicate an increasing demand sensitivity to prices. It is premature, however, to assess to what degree declining consumption was a response to high prices or a function of economic recession and stagnation. Petroleum remains the primary fuel consumed by the Western economies, a role it will continue to hold for years to come. There remains, as yet, no viable synthetic or natural liquid substitute for petroleum. Such alternative fuels, moreover, even if developed, could not be produced on a large scale for many years and probably would cost more than the prevailing price of oil.

OPEC, however, is not a homogeneous body that speaks with unity of voice or purpose. At best, it maintains a precariously balanced harmony of 13 voices; at times it is a noisy cacophony of disagreement. As is the case with any organization, the interests of the members frequently diverge, and individual states tend to put their interests ahead of those of the organization or of the majority of its members. Like other intergovernment organizations, OPEC is comprised of sovereign states, each of which recognizes no higher political authority.

The disagreements and conflicts between members both transcend petroleum issues—as in the case of the Iran-Iraq War, the political rift between Saudi Arabia and Libya, or the territorial disputes involving Iran, Iraq, and Kuwait—and are centered on oil policy. The issue that has been a chronic thorn in the organization's side has been the coordination of production. In simplest terms, OPEC has proved unable to agree on the absolute volume or share of the group's volume to be lifted by member countries. The allocation of market shares in March 1982 was an emergency measure meant as a short-term solution. Some members, notably Iran, refused to be bound by the OPEC policy of holding firm on prices while trimming production and sold increased volumes of crude at discounted prices. Although the 1982 prorationing scheme was not ineffective, OPEC has yet to gain the measure of control over output that it needs to regulate the world oil market effectively.

The more divisive concern, and the one that receives the most publicity, is pricing. The organization appears resigned to its inability to determine production. It has continuously sought, however, to maintain a coherent pricing structure in which the range of prices by members is limited and logical. Following the onset of the Iranian revolution, there was a continuous disintegration of OPEC's pricing policy. As some states constantly pushed their prices higher, the span between the highest and lowest prices exceeded $15 per barrel. By a combination of raising official prices and adding premiums, crude prices of the North African producers, for example, climbed beyond $40 per barrel, while Saudi marker crude was set at $26 per barrel.

It took more than two years—during which OPEC experienced a Saudi-orchestrated glut on world crude markets, cuts in production and prices by a number of countries, and declining purchases by the West—until the members could agree on a new pricing structure. With Saudi marker crude pegged at $34 per barrel, official prices range from $32 to $38 per barrel. The more aggressive price hawks among the group—most notably Libya, Algeria, Nigeria, and Kuwait—are not happy with the current pricing arrangement, which would not have been realized if not for the purposeful "overproduction" by the Saudis. Given opportune market conditions, the countries will no doubt jump at the chance to raise prices, thus once again disrupting the pricing structure.

The resolution of OPEC as a unified group was displayed somewhat by the efforts to force the oil companies to continue the lifting of expensive Nigerian crude in the spring of 1982. With world prices falling, the companies wanted Nigeria to roll back prices or release them from their obligations to produce from Nigerian fields. Following the March 1982 OPEC session, at which Nigeria's production quota was set at 1.3 million barrels per day, the country's output plummeted from 1.2 million barrels per day to less than 650,000 barrels per day. Given Nigeria's overwhelming dependence on the petroleum industry, which accounts for more than 90 percent of national exports, and the country's large population (being the most populous state in Africa), ambitious development plans (including the construction of a new capital), and foreign debt problems, the Nigerian economy was headed for disaster. If Nigeria's will or treasury was broken, moreover, the pressure on other OPEC countries would increase.

Once again the Saudis stepped in and assumed the leadership role. The Saudis warned that any company that reduced its liftings in Nigeria might be the target of sanctions by the Gulf producers and possibly by all of the OPEC countries. Fortunately for both sides, the showdown never took place. Prices had begun to rise as the supply overhang decreased and the market became more firm. Perhaps partially chastened by the Saudi warning, but certainly responding to changing market conditions, the companies operating in Nigeria increased their liftings in the African country to almost 1 million barrels per day before the end of April 1982 and to approximately 1.25 million barrels per day the following month.

The complexion of OPEC should undergo a major change in the medium term. The importance of the Persian Gulf states will grow as production by the North African and Latin American countries begins to wane. In the short term, Ecuador, Gabon, and Indonesia are expected to experience declines in output. Shortly thereafter, over the course of the 1990s, liftings in Venezuela, Algeria, and Nigeria may fall significantly. The Persian Gulf countries, which have the largest economic surpluses and therefore the least incentive to keep production high, may account for as much as 90 percent of OPEC exports by the year 2000, as opposed to the two-third share they held in 1980.[15] If such a pattern does emerge, the organization's power and unity may become greater. On the other hand, other leading Third World crude producers—particularly Angola and Mexico—probably will join OPEC, introducing new heterogeneity and diluting the group's concentration in the Persian Gulf.

An Organization of Gas Exporting Countries?

OPEC remains the archetypical commodity-producer group. It has inspired similar efforts, with varying degrees of success, by producers of bauxite, copper, iron ore, mercury, lead, and other mineral commodities. Without exception, none of the other commodity-producer groups has enjoyed the successes realized by OPEC. These groups invariably lack one or more of the important attributes we previously mentioned as central to OPEC's effectiveness.

Led by Algeria, there is a growing interest in the formation of a natural gas producer group, which we shall call OGEC—the Organization of Gas Exporting Countries. As an analog to OPEC, OGEC would represent the joint interests of Third World exporters of natural gas (pipeline and LNG). Other OPEC gas producers, including Libya, Nigeria, Qatar, and Iran, also have expressed interest in creating an OGEC.

Despite a history of production as long as that for petroleum (since crude and gas are often coproduced), the natural gas industry has yet to attain maturity. Until recently, natural gas was considered a nuisance, and most of it was flared at the well site, a practice that is declining but still widespread. The technologies for pipelining and liquefying gas, which are essential for trade in the commodity, are new and have yet to be fully developed. More than 85 percent of global commercial natural gas production is consumed in the country of origin—approximately twice the share of crude used in the country of production. World trade in natural gas in 1980 was just over 7 trillion cubic feet (200 billion cubic meters), the equivalent of barely 3.3 million barrels per day of crude.

Petroleum remains the primary hydrocarbon fuel, with natural gas traditionally treated as a secondary by-product. To a large extent, grouping petroleum and gas makes sense conceptually and in terms of geology, economics, and politics. Crude and natural gas often are found in conjunction and are coproduced, permitting the sharing of a significant percentage of fixed costs. Countries with large petroleum reserves tend to have large natural gas resources. Coherent policy necessitates coordination in developing both hydrocarbon resources. The national oil companies usually exercise authority with respect to both petroleum and gas.

Treating the two commodities simultaneously, however, invariably means subsuming natural gas concerns under the umbrella of petroleum issues. This is increasingly unacceptable to those countries that are trying to push natural gas prices to parity with crude. Moreover, the cast of leading actors within OPEC differs on oil and gas issues. Saudi Arabia, the leading OPEC oil producer, did not export any natural gas in 1980, nor did Kuwait, Iran, or Qatar. Abu Dhabi was the leading

Middle Eastern exporter of natural gas in 1980. Algeria and Indonesia, which accounted for less than 4 percent and 6 percent of OPEC oil sales in 1980, respectively, are the leading OPEC exporters of natural gas. Differences in technology, transportation, and financing also distinguish natural gas from petroleum.

The complexion of the world natural gas industry is changing too rapidly for the interests of gas exporters to be accommodated under the mantle of OPEC. Global gas trade is expected to double during the 1980s, including a quadrupling of LNG shipments. Planned pipeline and LNG projects abound, as a number of states—including Nigeria, Saudi Arabia, Kuwait, and Qatar—have announced their intentions to begin gas exports in the 1980s. The OPEC countries with gas potential and limited reserve/production ratios in crude—such as Nigeria, Algeria, and Indonesia—can be expected to concentrate their efforts increasingly on natural gas. The absence of an OGEC will hinder their efforts and introduce a potential new area of divisiveness among OPEC members.

The prospects for an effective OGEC are not promising in the short term. OPEC countries accounted for only 6 percent of world natural gas production in 1980 (down from over 7 percent in 1979), while the Third World commanded less than 13 percent of the total. OPEC members were the source of 12 percent of 1980 world exports and 17 percent of noncommunist exports. These figures compare poorly with the group's position with respect to world commerce in crude. In terms of LNG sales, however, OPEC members (Abu Dhabi, Algeria, Indonesia, and Libya) were the source of 72 percent of 1980 exports, with an additional 24 percent accounted for by Brunei, another developing country.

Despite their domination of world LNG markets, repeated efforts by Algeria, Indonesia, and Libya to obtain price parity with crude have failed. In its pricing battles with the United States and France, Algeria not only was unable to force price parity on consumers but suffered from a drop of almost 50 percent in its gas exports (all LNG) in 1980. Libyan efforts also culminated in decreased sales abroad.

Unlike the petroleum market, the leading exporters of natural gas are developed countries. The Soviet Union was the primary source of natural gas on world markets in 1980, accounting for more than 28 percent of total exports. Almost half of Soviet gas exports are destined for Western Europe. Soviet sales to this area should grow rapidly during the 1980s as the Yamal pipeline brings increasing volumes to the region. The quantity and price of Soviet gas (which is reported to be at least $2 below the approximately $7 per million BTU that is necessary for parity with crude) will continue to limit the effectiveness of an OGEC in manipulating the market. The next-largest exporters of natural gas are the Netherlands (the leading supplier of Western Europe), Norway, and Canada.

The share of trade handled by the four leading exporters will slip throughout the 1980s as exports from the Third World grow. The ceiling placed by reserves on OPEC's potential share of natural gas production, however, is considerably lower than that for crude oil. The OPEC members possess less than one-third of world reserves, with 40 percent of that amount in Iran. Even if an OGEC included all the OPEC states as well as Mexico, Brunei, Malaysia, Argentina, China, and Pakistan—an unlikely assortment of cooperating countries—it would hold less than 40 percent of world reserves. OPEC doesn't, however, have 55 percent of noncommunist gas reserves, and our hypothetical OGEC has about two-thirds of the total.

Without inviting the developing countries to participate—an act for which there is no precedent, except the special case of Australian membership in the International Bauxite Association—an OGEC would not be the powerful market force that OPEC is. Over time, as the market matures and as its long-term structure becomes more established, an OGEC may prove able to exercise a significant measure of influence on the relationship between the Western firms and the host governments. The absence of an OGEC does not mean, of course, the absence of political risks. To the contrary, the world gas industry is rife with political risks. The 1980 production cuts by Algeria and Libya, the escalating pricing demands by LNG exporters, the changes in the investment regimes regulating companies that produce natural gas in the Netherlands and Norway, and the prospects of increased reliance on Soviet supplies are but a few examples of the types and assortment of risks that are present in the world gas industry.

Notes

1. Raymond Vernon, "Introduction," in *The Oil Crisis* , ed. Raymond Vernon (New York: Norton, 1976), p. 5.
2. *Petroleum Economist*, June 1981, p. 257.
3. *Petroleum Economist*, August 1980, p. 329.
4. Ibid.
5. Zuhayr Mikdashi, *The International Politics of Natural Resources*, (Ithaca: Cornell University Press, 1976), p. 157.
6. *Petroleum Economist*, August 1980, p. 330.
7. *Petroleum Economist*, August 1981, p. 327.
8. *Petroleum Economist*, December 1979, p.529.
9. "Editorial," *Oil and Gas Journal*, 26 October 1981, p. 57.
10. *Wall Street Journal*, 10 December 1980.
11. *Petroleum Economist*, August 1980, p. 327.
12. *Financial Times*, 5 November 1980.
13. Gordon H. Barrows, "Special Report on World Petroleum Concessions," *Petroleum Economist* (October 1980): 426, 428.
14. Ibid., p. 426.
15. *Wall Street Journal*, 30 September 1981.

4

Risks and Resources in Oil and Gas

Reserves and Production

The distinction between reserves and resources is conceptual, technological, economic, and temporal. Reserves are those quantities of crude or natural gas that are economically recoverable, given prevailing technology and market conditions. The distinction is operational. Over time, although previous production has depleted a share of reserves, improved economic conditions and superior technology might swell the amount of proved reserves. Resources of natural gas and crude, on the other hand, may be four or five times as large as exploitable reserves, providing an important bank for future exploitation.

Reserves are an approximate indicator of the amount of oil and gas that can be produced under current conditions. Changes in prices, technology, synthetics, and other related areas provide a dynamic framework within which reserves are calculated. It is impossible to measure future reserves. Moreover, many countries, particularly those in the Third World, are underexplored, leaving gaps in our knowledge of the reserve/resource base.

Table 4.1 summarizes world production and reserves of crude oil and natural gas. There appears to be little correlation between reserves and output. In both oil and gas, the OPEC and non-OPEC Third World states command a far larger share of reserves (total and noncommunist) than of production. This is because of the uneven pattern of exploration and investment in the world industry. Despite the surfeit of reserves (and additional resources) in developing countries, the investment

Table 4.1 World Crude Oil and Natural Gas Production, 1980, and Estimated Reserves

	Petroleum		Natural Gas	
	Production (mb/d)	Reserves (bn bbls)	Production (bn ft³)	Reserves (bn ft³)
Canada	1.7[a]	8.0[a]	2,404	87,978
United States	10.2[a]	32.0[a]	20,084	200,175
Australia	0.4	2.5	338	30,010
Netherlands	0.0	0.0	3,080	56,345
Norway	0.5	5.5	890	46,390
United Kingdom	1.7	14.8	1,315	25,348
Other OECD	0.5	3.2	1,612	25,487
Total OECD	15.0	66.0	29,723	471,733
Algeria	1.1[a]	8.3[a]	353	131,473
Ecuador	0.2	1.0	3	4,060
Gabon	0.2	1.0	2	495
Indonesia	1.6	9.5	1,015	27,008
Iran	1.6	58.0	292	388,347
Iraq	2.7	30.1	62	27,537
Kuwait[b]	1.6	66.0	117	33,115
Libya	1.8	23.0	120	23,795
Nigeria	2.1	16.5	71	40,990
Qutar	0.5	3.5	109	65,240
Saudi Arabia[b]	9.9	166.0	525	94,545
United Arab Emirates	1.7	30.4	230	22,433
Venezuela	2.2	18.0	310	44,130
Total OPEC	27.2	431.3	3,209	903,168
Angola	0.2	1.3	9	1,060
Argentina	0.5	2.5	348	21,995
Brunei	0.2	1.7	315	7,695
Egypt	0.6	3.1	77	2,965
India	0.2	2.6	50	9,532
Malaysia	0.3	3.0	4	30,010
Mexico	2.1[a]	47.0[a]	1,142	77,495
Oman	0.3	2.9	21	5,013
Other non-OPEC Third World	1.4	6.9	1,330	93,737
Total non-OPEC Third World	5.8	71.0	3,296	249,502
China	2.1	20.0	503	25,950
Soviet Union	12.0	63.0	15,357	1,076,780
Other Communist	0.4	3.0	1,702	17,525
Total Communist	14.5	86.0	17,562	1,120,255
Total World	62.5	654.3	53,790	2,744,658

[a] Includes natural gas liquids.
[b] Includes share of production/reserves from Saudi-Kuwait Neutral Zone.
SOURCE: Compiled from various sources.

emphasis is on the developed states, which on the whole have the needed infrastructure, skills, and domestic market and are considered politically safer.

The OPEC countries also command a huge shut-in capacity. Having peaked at 31.4 million barrels per day in 1977, OPEC output has steadily fallen in recent years. During the first months of 1982, combined production for the organization was approximately 21 million barrels per day, indicating a vast reserve capacity that could be brought online rapidly. Following the March 1982 OPEC session, output dropped to less than 18 million barrels per day, the lowest level since the mid-1960s. Led by Saudi Arabia, which commands shut-in capacity of more than 3 million barrels per day (as shown by the country's record liftings in 1981), OPEC controls more than 10 million barrels per day of reserve capacity. This contributes further to the organization's market leverage.

Unless the reserve/resource base expands at a rate similar to production—an unlikely scenario—or substitutes become readily available, future production will become increasingly dependent on tapping the hydrocarbon reserves of the Third World. The OECD countries have a petroleum reserve/production horizon of 12 years, compared to 16 years for the communist countries, and more than 30 years for both OPEC and other Third World nations. At the extremes, the United States has petroleum reserves barely sufficient to maintain current rates of production for eight years, while Kuwait can continue to lift at 1980 levels for more than 100 years. World reserves are sufficient to accommodate current production for approximately 28 years.

The reserve/production ratios must be understood in context. The United States has had reserves equal to eight years of production for over a decade and may well have the same time horizon on reserves in 1990. Exploration is an ongoing activity, finding new reserves while old ones are being consumed. At a point, there is only minimal motivation for exploration. With such vast reserves relative to production, Kuwait has little incentive to invest the capital to prove additional reserves. Reserve profiles, therefore, are only general indicators of potential future production and must be evaluated in light of the array of variables affecting exploration, production, and profitability.

The reserves picture with respect to natural gas appears more optimistic than that for petroleum, with world gas reserves sufficient to satisfy 1980 production for 50 years. Again at the extremes, the United States has 10 years' worth of proved reserves, compared to more than 1000 years' worth in Iran. The market is too young, however, for this reserve/production profile to have much meaning. What the data do indicate is a vast natural gas potential that is not nearly so developed as the petroleum potential. As in petroleum and other minerals, developing countries, because of their lower demand for such commodities, have a surplus of natural gas reserves far in excess of their domestic needs. This can be an important source of leverage over world markets.

The market strength of Third World oil and gas exporters is bolstered by the resource-poor developed countries, particularly Japan. Japan is virtually 100 percent dependent on imported energy. It is not surprising, therefore, that Japan was readily willing to accommodate the political demands of the Arab oil-exporting countries during the 1973–1974 oil embargo and to accede to price increases on LNG imports from Indonesia and Brunei in the early 1980s.

Exploration

Exploration is a key variable in the evolution of the industry; it is an indication of priorities and expectations, measuring the future against the past. In a market free of political influences, exploration expenditures would be allocated proportionally according to the volume of reserves and delivered costs from the various export points. The contemporary petroleum and natural gas market is anything but apolitical. The pattern of exploration activities reveals not only the economic and geological logic of the industry but is inseparable from the web of political issues centered on the fuels.

The overt efforts to diversify sources and minimize reliance on OPEC has stimulated renewed exploration efforts in non-OPEC countries. Domestic explorations in the United States and Western Europe were at record levels throughout 1981, with the OECD countries the recipients of some 90 percent of exploration monies spent in search of crude and natural gas. Not only are domestic supplies seen as the most safe politically, but the governments in a number of OECD countries have offered political inducements in the form of tax breaks, price increases, or subsidies to spur exploration. These are evidence of the new opportunities that may become available to firms as a result of political changes and decisions.

Diversification of supply is a standard strategy in minimizing the risks from overreliance on a particular state or group of states. Western interest and investment in new non-OPEC explorations abound, spanning the globe from Argentina to Brazil, Madagascar to Sudan, and the Philippines to China. The terms of agreements in non-OPEC developing countries often are more inviting and rewarding for foreign firms than the opportunities in OPEC states. The non-OPEC nations, moreover, are more likely to lack the capital, expertise, and other attributes that contribute to the bargaining strength of the OPEC members. This not only affords more favorable terms to foreign firms but also may protect the oil companies from rapid changes in agreements in favor of the host states.

Exploration is an area of expertise in which the oil firms control vital skills and abilities and are prepared to assume the financial risks if the rewards are sufficient. The less-experienced nocs often are incapable of executing thorough exploration programs, and most Third World governments are reluctant to bear the onus of the high risks that are endemic to oil and gas exploration. Most countries continue to rely on the transnational oil companies to carry out exploration, often in joint ventures with the nocs. Peru and Brazil, for example—both of which had oil industries that were stagnating for lack of investment and were confronted with declining reserves—recently offered more inviting terms to induce investment in new exploration projects. In 1980 President Figuerdo cleared the political air that had deterred exploration activities in Brazil for a number of years by confirming his belief in the private sector. Similarly, in December 1980 the Belaunde government relaxed the provisions governing petroleum exploration and development in Peru and offered tax credits of 40 percent to 50 percent on income reinvested in the country's oil industry.

Declining efforts by the transnational firms to explore for additional reserves in the OPEC countries are reflected in the decrease, however slight, in the amount of reserves held by these states. As a group, OPEC reserves have fallen about 1 percent since 1975. Algeria and Saudi Arabia (which relies on extensive involvement by the Aramco partners) were the only OPEC countries that increased their reserves (approximately 10 percent for each) in the 1975-1980 period. Without the growth in Saudi reserves of approximately 15 billion barrels, OPEC reserves would have slipped more than 4 percent in the latter half of the 1970s. In relative terms, reserves in Qatar declined 40 percent, Ecuador 33 percent, Nigeria 18 percent, Indonesia 14 percent, Libya 11 percent, and the UAE 5 percent. The remaining members had unchanged reserves or experienced downfalls of less than 5 percent. For the same period, reserves in the remainder of the world climbed 13 percent, including a jump of more than 150 percent in non-OPEC Third World countries.

The drop in reserves among the OPEC states in the 1975-1980 years contrasts sharply with the growth in reserves from 1970 through 1975, when the transnational oil firms were more active in exploration projects in member countries. OPEC reserves climbed more than 21 percent in the first half of the decade, as every member had larger reserves in 1975 than in 1970. In the 1970-1975 period, crude reserves grew 150 percent in the UAE, doubled in Nigeria and Ecuador, and increased by 30 percent in Qatar, Venezuela, and Gabon.

There is a growing concern among the OPEC nations about the slip in reserves and the decrease in exploration activities. Net additions to Middle East petroleum reserves, according to Dr. Fadhil J. Al-Chalabi, OPEC's deputy secretary-general, were only one-quarter of cumulative production during the 1970s, while production totaled only a small fraction of net additions in the 1960s. To a large extent, this situation is the result of the "meager investment effort made in the OPEC countries in general and the Middle East Arab producing countries in particular."[1] In the 1973-1979 period, the number of wells drilled in the Middle East declined 10 percent, whereas increasing prices should have stimulated increased investment in exploration, as they did in other countries. Although it may not be showing on the surface, the OPEC countries need the involvement of the transnational oil companies in exploration activities.

Algeria and Libya have linked prices and access to supply to investments in exploration in an effort to prove additional reserves. Indonesia and Nigeria continue to involve foreign oil firms as vital partners in exploration efforts, with the risk borne by the companies. Even Iraq, which has long been hostile to Western capital and influence, has sought the assistance of the transnational oil firms in survey and exploration work. The OPEC countries seem to recognize, albeit reluctantly, the importance of the Western oil companies. As exploration investments continue to lag in the OPEC states, the role of the firms in global exploration will become more pronounced.

In other areas as well, the oil TNCs remain important in the national oil industries of the OPEC countries. Despite the trend toward a reduced corporate presence and host government wishes to be free of the foreign firms, the exploration skills of the companies are in demand. Similarly, the managerial abilities of the oil companies remain important. The continuing role of Aramco as manager of Saudi operations and the renewal of oil company service contracts are evidence of the continuing role of the companies in oil development projects. Moreover, Corporacien Venezolana del Petroleo (CVP) recently invited foreign oil firms to bid on general contract and supervisory work for the development of the heavy oil fields. This is a sharp reversal in Venezuelan policy, which has denied the transnational oil companies any role in oil and gas projects for the last six years.

U.S. Direct Investments Abroad

Except for those countries that have totally nationalized their oil and gas industries, direct investment in hydrocarbons is still the norm. Usually, the foreign firm is a minority participant in a project, but direct equity investments in oil and gas are permitted, with varying degrees of restrictions, in all but a handful of countries, such as Iraq, Kuwait, and Mexico.

In many respects, ownership is a secondary issue. Profits can be made and supply obtained regardless of ownership. Equity ownership, moreover, presents additional risks to the firm in the form of the value of assets in place. A number of countries that lack the surplus capital to buy out the foreign firms completely are comfortable with a simple majority share in the hands of the host government. The government is able to control operations and decisions and net huge profits regardless of the foreign equity position and enjoys the advantage of employing foreign capital in the place of what often is scarce domestic capital.

Incidents of forced divestiture seem to have declined in the latter half of the 1970s,[2] following a period during which expropriation, nationalization, and participation struck virtually every petroleum and natural gas project in the Third

World, as well as a number of projects in the OECD states. This trend has been particularly relevant in the Third World, where the 1972 level of U.S. foreign investment in petroleum was not surpassed again until 1980. The data on the book value of U.S. direct investments in petroleum operations are contained in table 4.2.

Over the past decade, total U.S. direct foreign investments in oil have grown 115 percent, or at an annual rate of less than 8 percent. Investments in the developed countries have increased more than 10 percent yearly, growing 170 percent between 1971 and 1980. The main OECD recipient of U.S. investment capital is Canada, which held 31 percent of the more than $34 billion invested in the oil industry in the industrialized states. Down from 37 percent at the opening of the decade, the amount and share of U.S. investment in the Canadian oil industry should drop substantially as the Trudeau government implements its Canadianization policies. The tripling of American oil assets in the United Kingdom and a tenfold increase in Norway since 1971 account for a large share of the growth in direct investments in the European countries.

In 1980, the developed countries held 73 percent of total U.S. direct petroleum investments abroad. This reflects the lag in such investments in the developing countries. The position of U.S. oil firms in the Third World grew by 46 percent in the 1971–1980 period, an annual growth rate of just over 4 percent. In 1971, the Western nations were the target of 58 percent of U.S. direct investments in oil. The increased share of investment in the developed countries reflects the differences in perceived risks between the developed and developing world and the wholesale forced divestment the oil firms suffered in the Third World during the mid-1970s.

Forced divestments of U.S. oil assets in the Third World, particularly in the Middle East, prompted a devastating decline in the U.S. investment position in such countries in 1973 and 1974. From a record level of $47.4 billion in 1972, U.S. direct oil investments in the Third World plummeted to – $390 million in 1974. The position in non-Latin American Third World nations ran a continual deficit from 1974 through 1978, as forced divestments exceeded the reported book value of holdings in those countries. Oil investments in the Middle East, for example, fell from a reported level of $2.1 billion in 1973 to – $4.8 billion in 1976. Although direct petroleum investments in Latin America have fared somewhat better, the U.S. position in Venezuela has slipped from $1.7 billion in 1971 to a mere $3.9 million in 1980.

Since 1978, U.S. petroleum investments abroad have increased an impressive 54 percent, growing by 35 percent in the developed countries and by almost 300 percent in the Third World. Only $1.3 billion of the $7.6 billion increase in the oil position in developing countries has been concentrated in Latin America. The growth in petroleum investments abroad in 1979 and 1980 (with similar results revealed by preliminary 1981 data) indicated increased interest in developing additional petroleum reserves in both the Third World and the West.

Table 4.2 U.S. Direct Investments Abroad in Petroleum, 1971–1980 (millions of dollars, year-end)

	1971	1972	1973	1974	1975	1976	1977	1978	1979	1980
Developed countries	12,544	13,542	15,911	18,204	20,129	22,974	24,854	25,341	30,220	34,173
Canada	4,643	4,764	5,320	5,731	6,220	7,181	7,722	7,686	8,648	10,573
Europe	6,247	6,872	8,524	9,830	11,165	12,726	13,926	14,326	17,755	19,924
Other	1,654	1,906	2,066	2,642	2,744	3,068	3,206	3,329	3,816	3,676
Developing countries	7,027	7,376	6,074	(−390)	2,519	2,261	3,014	2,648	6,093	10,271
Latin America	2,939	2,979	3,043	3,564	3,324	2,932	3,378	3,088	3,948	4,336
Other	4,088	4,397	3,030	(−3,954)	(−805)	(−671)	(−365)	(−440)	2,145	5,935
TOTAL[a]	21,794	23,385	24,951	21,418	25,972	28,408	30,887	30,532	38,744	46,920
(as % of all U.S. foreign direct investments)	(26.3)	(26.0)	(24.6)	(19.5)	(20.9)	(20.8)	(20.8)	(18.8)	(20.7)	(22.0)

[a] The differences between the sums of developed and developing countries and yearly totals result from assets described as "international and unallocated."

SOURCE: United States Department of Commerce.

Supply

A central concern of the transnational oil firms and their home governments is assured access to sufficient supplies of crude oil and natural gas. Policy decisions increasingly impinge on supply decisions. Access has been linked to a variety of conditions, including associated investments, purchase of refined products, limitations of destination, supply of technical services, use of national flag tankers, and other demands.

The historical problem of oversupply, with which the major companies struggled to avoid weakening prices amid continuous market gluts, has yielded to a tight market in which shortage is more common than plenty. Except for the surplus crude available for most of 1981 and 1982, the contemporary industry suffers from a chronic problem of scarcity and potential shortages. The year 1981 was peculiar in that there were a number of instances in which the companies refused opportunities to purchase crude: Ashland Oil suspended purchases of crude under contract from Cameroon and Mexico, Atlantic Richfield cut purchases from Nigeria, and CFP reduced by half its imports from Mexico. There also were reports that in Libya, for example, firms were reducing liftings under the guise of technical maintenance work on producing wells.[3] By withdrawing from Libya in 1981, Exxon voluntarily relinquished its right to some 65,000 barrels per day equity crude and probably will find it difficult to continue to purchase additional volumes of Libyan oil under contract. Similarly, throughout 1982 Nigerian liftings fell by more than half, as the operating companies turned away from high-priced Nigerian crude in favor of less-expensive sources of comparable quality.

Aside from exceptions to the rule, which have been mentioned, the oil companies are always anxious about obtaining sufficient supply and are interested in additional sources and volumes of crude. Equity liftings and sales under contract are the prerogative of the host government, leaving the companies vulnerable to changes in policy. Until 1980, the foreign firms with equity interests in Nigerian projects, for example—which include Mobil, Shell, Texaco, Phillips, and Elf—were entitled to 40 percent of their *allowable maximum* respective liftings (80 percent in the case of Shell). The Frikefe Tribunal, however, decided that the firm's share should be based on *actual* production. As a consequence, not only did the companies have to compensate the Nigerian National Petroleum Company (NNPC) for the excess quantities of crude they had exported in earlier years, but the quantities of equity oil to which they are entitled was pegged to production levels. Throughout 1981, this meant reduced and fluctuating volumes for the oil firms, as high-priced crude sales by NNPC were erratic. Nigerian production slipped from 2 million barrels per day in January, to barely 700,000 barrels per day in August and climbed to 1.3 million barrels per day before the end of the year.

In many respects, the decision of the Frikefe Tribunal was unexpected by both the companies and the government of President Shagani. The larger liftings by the foreign equity producers meant higher earnings for the government, which are particularly important when sales by NNPC decline. There are a number of more blatant examples of host governments' abruptly terminating or reducing liftings by foreign firms. The Qadhafi government, for example, mandated production cuts across the board for the foreign producers in Libya in 1980. In March of that year, the government announced production cuts totaling 350,000 barrels per day, some 17 percent of the country's output, distributed as follows: '

Occidental	down 36.7%	(a loss of 110,000 b/d)
Elf Aquitaine	down 28.6%	(a loss of 4,000 b/d)
Esso	down 27.0%	(a loss of 50,000 b/d)
Agip	down 20.0%	(a loss of 40,000 b/d)
Oasis Group	down 16.0%	(a loss of 120,000 b/d)
Mobil	down 10.0%	(a loss of 9,000 b/d)
National Oil Corp.	down 3.6%	(a loss of 20,000 b/d)[4]

In an even more unexpected move, Pertamina instructed Atlantic Richfield to halt production in the Kalimantan region of Borneo the day after the expiration of their ten-year production-sharing agreement. Pertamina justified its decision, the first of its type in Indonesia, on the grounds of unsatisfactory performance by the firm in terms of crude liftings, exploration, and additional investments.[5]

Limitations on liftings in the British North Sea were considered in 1982. This was the first year that the government could restrict production, as part of its oil depletion policy, on fields discovered before 1975.[6] The government was reluctant to limit output because of the adverse political effect such a decision would have on future explorations and investment. The British economy, moreover, is enjoying a much-needed boost from the rapid growth in oil earnings from exports.

At times, price and supply questions have been inseparably intertwined. Price squabbles have prompted a number of instances in which host governments have slowed or terminated supplies to purchasing companies; Algerian LNG to Gaz de France, El Paso, and ENI and Iranian crude sales to Japan are but a few examples.

Not all supply shortfalls are the direct result of purposeful government policies. The political volatility of the Persian Gulf and the vulnerability of the Straits of Hormuz—through which the majority of Persian Gulf crude flows—were painfully enunciated by the outbreak of the Iran-Iraq War in the fall of 1980. Almost overnight, more than 15 percent of OPEC exports—700,000 barrels per day of crude and 300,000 barrels per day of products from Iran and 3 million barrels per day of crude from Iraq—evaporated from world markets. Besides contributing to a doubling of prices, the conflagration threatened to spread; Kuwait reported being subject to intermittent bombings, there was talk of blockading or mining the Gulf, and some Arab states were reportedly considering aiding Iraq.

The catastrophe did not happen. After initially targeting their opponents' petroleum installations for destruction, the belligerents seemed to arrive at a tacit *modus vivendi* by which they avoided direct assaults on each others' oil facilities. The war did not spread, nor did it block the flow of the area's oil and gas trade. The lingering doubts about the future stability of the area and the potential for a new, more-extensive outbreak of regional war is perhaps best confirmed by the current Saudi debate on whether to build a strategic oil reserve and by the recent completion of a 750-mile pipeline to a new export terminal at Yanbu on the Red Sea (at a cost of $1.6 billion).[7] Sheikh Yamani, Saudi minister of petroleum, already has speculated that the pipeline might be expanded or that a second line could be built. The pipeline reduces Saudi dependence on the Persian Gulf and the Straits of Hormuz, both of which are militarily vulnerable, while the strategic reserves would afford the country emergency supplies in the event of military action forcing the cessation of production.

Of course, there is also the perennial fear of a third embargo. The first oil embargo, imposed in 1967, proved ineffective. The OAPEC countries successfully embargoed selected countries, including the United States, the Netherlands, and Portugal, for a few months in 1973–1974. The embargo is seen as an important policy tool by the OPEC countries. Selective embargoes are maintained against Israel and the Republic of South Africa, while *threats* of embargoes have been invoked against the United States and Western Europe. In 1981, for example, Libya's Union of Petroleum, Mining and Chemical Workers called on the Arab oil exporters to embargo the United States in protest to the increased American military presence in the area surrounding the Persian Gulf and the Mediterranean Sea,[8] while Iraq recommended a cut in crude sales to the United States and to those Western countries that are stockpiling oil.[9]

The greatest risk of a renewed embargo, however, centers on the motive behind the previous embargoes: U.S. support for Israel. In 1980, Algeria and Libya agreed to prohibit the sale of oil and gas to countries supporting Israel,[10] although no such policy has yet to be implemented. Israeli annexation of the Golan Heights—the course of events surrounding which have yet to unfold fully—prompted new efforts to embargo oil sales to the United States. Syrian President Assad lobbied heavily in the Middle East OPEC countries, particularly Saudi Arabia, in his efforts to mount a unified front against Israel and her main supporter, the United States. The Israeli attacks on PLO installations and occupation of parts of Lebanon during the summer of 1982 and renewed fighting with Syria once again raised the specter of an embargo.

An embargo may not mean reduced liftings for or sales to the oil companies, although the 1973–1974 action was combined with production cuts that reduced the availability of supplies. The firms might once again be caught, however, in a political crunch, with pressures from both their home and host governments. The political risks surrounding the industry under such conditions multiply exponentially, increasing the likelihood of politically induced losses to the oil firms.

Trade Flows

The petroleum industry is geared to the world market. Oil is the most important commodity in global trade in terms of value, tanker miles, and tonnage shipped. Approximately one-fifth of the total value of international trade is accounted for by petroleum, and 85 percent of the oil trade is in crude. World trade in natural gas is still in its infancy. With the vast majority of gas consumed in the country of extraction (and large amounts flared rather than captured) and with most trade links only a few years old, the natural gas industry is considerably less oriented to world markets than its sister hydrocarbon. As the world gas industry matures, it will become increasingly international in structure, in a manner akin to petroleum.

By source, 62 percent of the approximately 26.5 million barrels per day of crude oil traded in 1980 originated in the Middle East. The North African producers accounted for 19 percent, while Latin America provided 7 percent of worldwide exports. By destination, Western Europe was the main market, receiving just under 40 percent of global crude imports. The United States and Japan were the next leading purchasers, accounting for 20 percent and 17 percent, respectively. Table 4.3 and map 4.1 summarize the data on world trade flows in crude and oil products.

Trade in petroleum products displays a different complexion. The leading exporter was Latin America, providing 38 percent of the 5.4 million barrels per day of oil products traded in 1980. The Middle Eastern countries were the source of just over 20 percent of product supplies, while North Africa contributed only 5 percent of the total. The United States was the largest consumer of product imports, receiving 28 percent, while Western Europe absorbed 27 percent of the total. The next leading products market is Japan, with just over 10 percent of purchases.

The United States relies on the Middle East and North Africa for 53 percent of total petroleum imports and for more than 60 percent of its crude imports. By region, however, Latin America is the largest U.S. source of petroleum. Western Europe receives 82 percent of its petroleum purchases from the Middle East and North Africa and approximately 85 percent of its crude imports. Japan obtains 93 percent of its oil imports and more than 95 percent of its crude imports from the Middle East and Southeast Asia.

The sourcing and destination of natural gas in world trade is markedly different from those in petroleum. Approximately 40 percent of the 7.1 trillion cubic feet of gas traded internationally in 1980 was exported from Western Europe, while an additional 28 percent was from the Soviet Union. The Middle East, North Africa, and

Map 4.1 Main Oil Movements by Sea, 1980

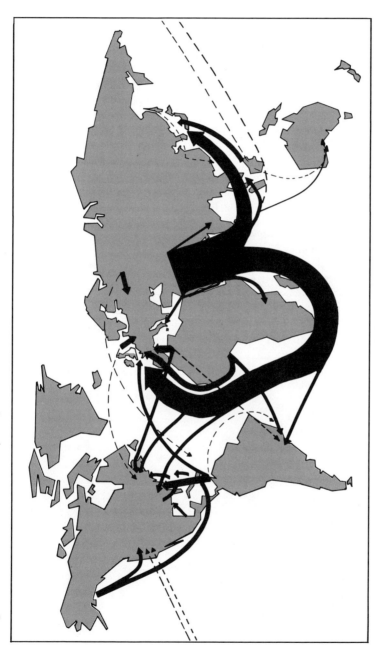

SOURCE: *BP Statistical Review of the World Oil Industry 1980* (British Petroleum, 1981).

Table 4.3   Regional Trade Flows in Crude and Oil Products, 1980
(thousand b/d)

Exports by \ Imports by	United States	Canada	Latin America	Western Europe	Africa	South-east Asia
United States	—	105	290	145	5	10
Canada	370	—	—	30	—	10
Latin America	2,050	215	275	625	50	—
Western Europe	355	5	—	—	85	5
Middle East	1,555	300	1,615	7,275	445	1,745
Africa	2,000	10	555	2,430	75	10
Southeast and South Asia	405	—	—	15	25	—
Japan	—	—	—	—	—	5
Australasia	—	—	—	5	—	10
USSR, E. Europe, China	—	—	205	1,300	—	225
TOTAL IMPORTS	6,735	635	2,940	11,825	685	2,020
CRUDE IMPORTS	5,220	595	2,630	10,390	510	1,635
PRODUCT IMPORTS	1,515	40	310	1,435	175	385

[a] Includes changes in quantity of oil in transit, transit losses, minor movements not otherwise shown, unidentified military use, etc.
[b] Includes 475,000 b/d crude imported by South Asian countries and 165,000 b/d by the USSR, Eastern Europe, and China.
[c] Includes 145,000 b/d products imported by South Asian countries and 195,000 b/d by the USSR, Eastern Europe, and China.

SOURCE: *BP Statistical Review of the World Oil Industry, 1980* (British Petroleum, 1981).

Japan	Austral-Asia	Other Eastern Hemisphere	Destination Unknown[a]	Total Exports	Crude Exports	Product Exports
—	—	—	—	555	285	270
10	—	—	25	445	200	245
80	5	235	350	3,885	1,840	2,045
—	—	10	—	460	300	160
3,560	220	695	100	17,510	16,420	1,090
70	—	155	—	5,305	5,030	275
1,080	140	80	—	1,745	1,350	395
—	—	—	—	5	—	5
5	—	—	—	20	—	20
180	—	95	—	2,005	1,120	885
4,985	365	1,270	475	31,935		
4,435	200	730[b]	200		26,545	
550	165	540[c]	275			5,390

Latin America, the three leading oil exporting regions, accounted for only 2 percent, 4 percent, and 2.5 percent of world natural gas exports, respectively. The leading market for gas, as for crude, is Western Europe, with 54 percent of total imports. Eastern Europe, the United States, and Japan were the other significant buyers, with the Eastern European countries accounting for 15 percent of purchases, the United States 13 percent, and Japan 12 percent.

The leading trade flows in gas, unlike in petroleum, are between industrialized countries. The OPEC and other Third World countries are not the primary sources of natural gas entering world trade. The intra-European trade is the most important regional market in the global gas industry, accounting for two-fifths of the total. (All Western European exports are to other Western European countries.) The most important bilateral trade flows in natural gas involve West German purchases of natural gas from the Netherlands, Norway, and the USSR; American imports from Canada; British acquisitions from Norway; and Belgian and French imports from the Netherlands. The only major trade flows (more than 150 billion cubic feet) involving the Third World are Japanese imports from Indonesia and Brunei (420 billion cubic feet and 275 billion cubic feet in 1980, respectively). Table 4.4 and map 4.2 present the data on world trade in natural gas by region.

The patterns of LNG and pipeline trade diverge widely. The Far East, the source of barely 11 percent of total natural gas exports, was the source of 62 percent of LNG entering world trade in 1980. North Africa contributed another 26 percent. As all of Western Europe's, the USSR's, and Canada's exports were via pipeline, these countries accounted for virtually 95 percent of world trade in pipeline gas. LNG purchase by Japan accounted for 100 percent of that country's gas purchases in 1980. The United States and Western Europe were the only other recipients of LNG, and in both instances LNG was less than 10 percent of gas imports.*

As in the case of petroleum, there is a high degree of reliance on LNG imports from particular regions. Unlike in oil, however, the United States and Western Europe are largely reliant on sources among other OECD countries: 82 percent of U.S. gas supplies come from Canada, while 71 percent of Western European imports are from other Western European states. Japan, on the other hand, is dependent on the Far East for 85 percent of gas imports and on the Middle East for another 11 percent. Western Europe also obtains 24 percent of its natural gas imports from the Soviet Union, a share that is scheduled to increase with the construction of the Yamal pipeline.

---

*Approximately 12 percent of French and 100 percent of Spanish gas imports in 1980, however, were LNG.

Table 4.4 Regional Trade Flows in Natural Gas, 1980 (bn)[3]

Exports by ╲ Imports by	United States	Latin America	Western Europe	Japan	USSR	Eastern Europe	TOTAL EXPORTS	LNG EXPORTS	PIPELINE EXPORTS
United States	—	3.5	—	42.4(a)	—	—	45.9	42.4	3.5
Canada	812.0	—	—	—	—	—	812.0	—	812.0
Latin America	102.4	77.7	—	—	—	—	180.1	—	180.1
Middle East	—	—	—	95.3(a)	24.7	—	120.0	95.3	24.7
Africa	74.1(a)	—	222.4(a)	—	—	—	296.5	296.5	—
Western Europe	—	—	2,863.2	—	—	—	2,863.2	—	2,863.2
Far East	—	—	—	695.5(a)	88.3	—	783.8	695.5	88.3
USSR	—	—	970.9	—	—	1,052.1	2,023.0	—	2,023.0
Eastern Europe	—	—	—	—	—	7.1	7.1	—	7.1
TOTAL IMPORTS	988.5	81.2	4,056.5	833.2	113.0	1,059.2	7,131.6		
LNG IMPORTS	74.1	—	222.4	833.2	—	—		1,129.7	
PIPELINE IMPORTS	914.4	81.2	3,834.1	—	113.0	1,059.2			6,001.9

(a) LNG.

SOURCE: Compiled from various issues of *Petroleum Economist*.

Map 4.2   Regional Trade Flows in Natural Gas, 1980

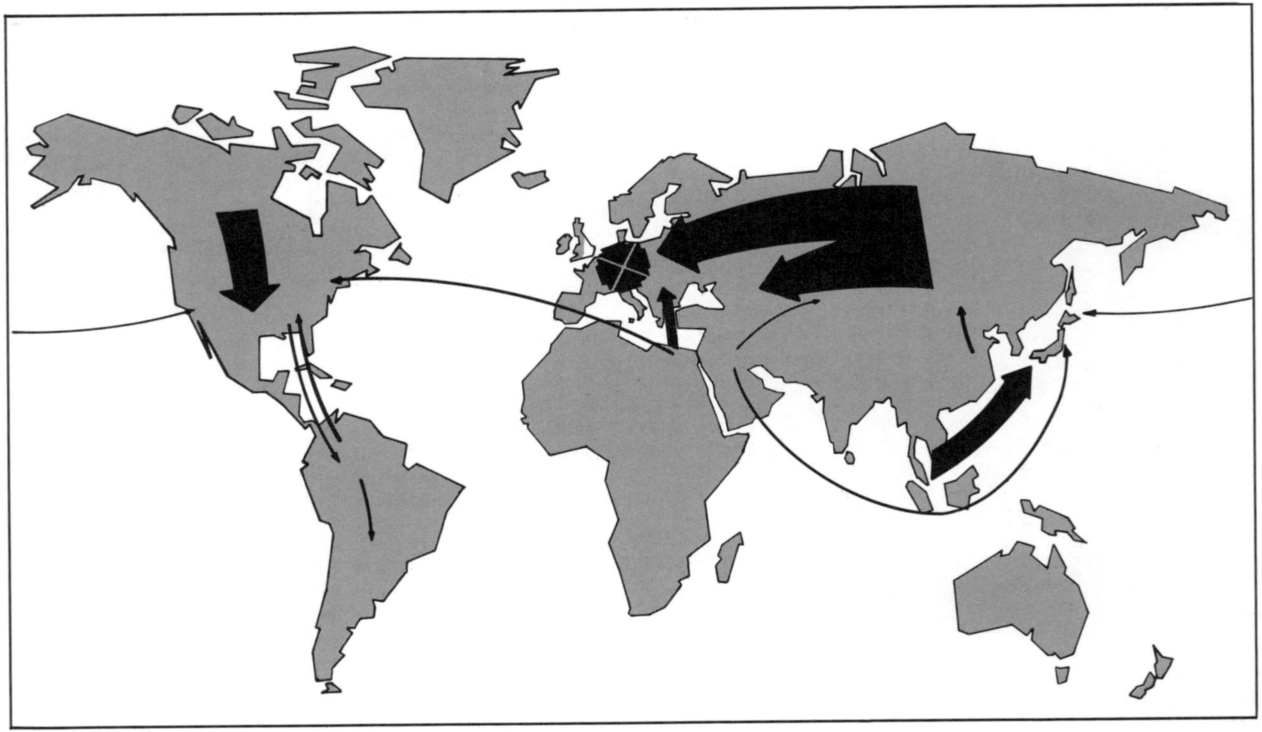

SOURCE: Based on data in table 4.4.

Oil and Gas as a Special Case

Investments in fuel and nonfuel minerals have a number of characteristics in common. In many respects, the risks are similar, as is the nature of projects. Both are capital intensive and involve long lead times to bring a new project onstream and lengthy payback periods to recover investments. Exploration is essential, costly, and risky. The natural resource industries, on the whole, have proved highly vulnerable to expropriation and nationalization. Net natural resource flows tend to be from Third World countries to the West. To varying degrees, multinational firms are important to the successive stages involved in natural resources projects, from exploration and development through processing, transporting, and marketing.

The similarities between other minerals and oil and gas, however, often are overstated. What sometimes is seen as a question of degree is substantively important to investments and to the structure of the world industry. Despite the general importance of natural resource exports to the economies of the Third World, petroleum is far and away the most valuable Third World export—exceeding the value of all other raw material exports from these countries combined. To treat the difference between oil and the hard minerals as one of mere magnitude, therefore, is to distort reality.

In a similar vein, oil and, to a lesser extent, natural gas, play a role in the world and national economies that cannot be rivaled by other natural resources. In both the literal and figurative senses, petroleum fuels the world economy. Natural resource firms tend to be big. Among such firms, however, the oil companies are giants. The 1981 *Fortune* 500 ranking, for example, lists 6 oil companies among the 10 largest U.S. firms and 13 in the top 20. In contrast, the leading minerals company in 1980, the Aluminum Company of America (Alcoa), is ranked number 63 on the list.

Trade in petroleum is more concentrated than that in hard minerals. Substitutes for oil and gas, moreover, are not readily available. Almost all hard minerals have substitutes, both natural and synthetic, to which consumers can turn if minerals producers make excessive demands. The vital nature of petroleum in the contemporary world, combined with the commodity's scarcity value, are the main ingredients in the bargaining power wielded by the oil-exporting countries and are the source of a good portion of the political risks with which the companies are confronted. The "prospectivity" of petroleum—the likelihood of finding commercial fields—and the importance of increasing supply often incline petroleum companies to understate political risks.

Natural resource firms have long been recognized as vulnerable to political risks. As the largest, most important, and most visible of such firms, the transnational oil companies have been the most frequent targets of nationalization and expropriation. One study reported that of 76 Third World countries surveyed, during the 1960–1976 period 54 countries forced the divestment of petroleum firms.[11] This was considerably higher than the number of countries that seized foreign assets in any other economic sector.

Petroleum is far easier to market than other minerals. This has enabled the host governments to negotiate direct sales abroad, bypassing the transnational oil companies. Marketing other raw materials has proved far more difficult for producing countries, especially in the case of highly integrated industries, such as the bauxite/aluminum industry. The economic and political inducements to direct export are strong: it increases the returns to the host country as well as affording the government additional control over the industry and the destination of exports. With minor exceptions, moreover, the oil-exporting countries can sell abroad without worrying about flooding the market or depressing prices. No other mineral commodities have a pricing structure that is as unilaterally controlled by the producing countries.

On the whole, the exploration risks are higher in hard rock minerals. Exploration costs, however, have been estimated to account for 5 percent of the total investment in bringing a new mine onstream and 25 percent of the expense incurred in bringing a new field into production.[12] Although both fuel and nonfuel mining industries tend to be highly capital intensive, oil companies traditionally have enjoyed a higher rate of return on capital. The higher rents associated with crude in comparison to hard minerals have often enabled oil companies to recapture investments in a shorter time span than that which must be endured by mining firms.

Oil-exporting Third World countries are far richer than the average developing nation. Increased oil prices and taxes and royalties have earned many of the host countries sufficient wealth to be able to afford participation in and nationalization of the national oil industry. Hard minerals-exporting developing countries, on the other hand, are almost universally capital-poor. These countries can ill afford to purchase an equity stake in the local mining industry. The high price of crude has meant that even the less-wealthy oil-exporting countries could afford to pay compensation or to cover the cost of equity acquired out of future production. Nigeria, for example is reportedly paying BP Naira £71.34 million ($135 million) worth of crude (at $40.02 bbl) for the 1979 nationalization of the firm's Nigerian production and marketing assets.[13] The value of crude is the key to the success host governments have enjoyed in attracting foreign firms to participate in production-sharing ventures and risk contracts.

Capital surpluses are unheard of among non-oil-exporting Third World countries, all of which suffer from chronic problems of insufficient hard currency earnings and supply. Oil sales guarantee a country hard currency income. The ability of a small, but central, group of oil-producing states—particularly Saudi Arabia and Kuwait—to curtail production and exports without risking their financial ruin is unique among the extractive industries and gives OPEC a greater degree of flexibility and stamina than other commodity-producer groups.

Although the differences between hard minerals and oil and gas should not be overstated, they command considerable attention. Petroleum and natural gas stand as exceptions to the general rules of the hard-rock-mining industries. The special attributes of these two hydrocarbons and their role in the world political economy render them unique. From this uniqueness stems a large measure of the political problems confronting the transnational companies.

Notes

1. *Petroleum Economist*, July 1981, p. 282.
2. Stephen J. Kobrin, "Political Assessment by International Firms: Models or Methodologies," *Journal of Policy Modeling* 3 (1981): 253.
3. *Wall Street Journal*, 23 June 1981.
4. *Petroleum Economist*, May 1980, p. 217.
5. *Wall Street Journal*, 16 October 1981.
6. *Financial Times*, 18 December 1981.
7. *Financial Times*, 18 November 1981.
8. *Wall Street Journal*, 16 November 1981.
9. *Wall Street Journal*, 25 November 1981.
10. *Petroleum Economist*, May 1980, p. 217.
11. Stephan J. Kobrin, "Foreign Enterprise and Forced Divestment in LDCs," *International Organization* 34 (Winter 1980): 78.
12. The Report of the Independent Commission on International Development Issues (Brandt Commission), *North-South: A Program for Survival* (Cambridge: MIT Press, 1980), p. 156.
13. *Petroleum Economist*, April 1981, p. 135.

5

The Nature of Political Risks

The Sources of Political Risk

When making a foreign investment decision the transnational company must evaluate the commercial feasibility of and expected returns from a project in light of the current investment climate and probable changes in the investment regimes. Initial conditions are known and measurable. The specific terms governing oil and gas ventures are contained in either the particular investment agreement or the relevant national legislation, depending on the approach of the host government.

From the firm's point of view, there is nothing inherently advantageous in having the host government rely on general petroleum/natural gas legislation or negotiate project-specific agreements. If the entire industry is subject to the same legislatively established conditions, the company can be more confident that it will not be singled out for discriminating treatment. Changes in legislation will affect all the foreign companies (although not necessarily to the same extent). The host government may be more reluctant to enact changes that antagonize all the foreign oil firms simultaneously, particularly since the companies' sense of shared fate encourages the formation of a unified front against the government.

The negotiation of separate corporate–host investment agreements, on the other hand, enables the transnational company to address the particular concerns of each project. This may be essential to bringing marginal fields online, completing projects that need extensive associated infrastructure, or drilling under adverse weather and environmental conditions. Moreover, the company can negotiate to the best of its abilities to exact more-favorable terms than are generally prevailing, although this tactic carries the corollary gamble of permitting the host government to negotiate to *its* best advantage and force the foreign firm to accept less-favorable terms.

As a rule, developed countries are more likely to employ general legislation in governing oil and gas investments, while developing states tend to rely on negotiated agreements. Growing institutionalization and bureaucratization in the oil-producing countries of the Third World are changing this pattern, and an increasing number of less-developed host countries are moving toward using legislation rather than in-dividual negotiations. Legislation does not insure a more favorable investment regime, however, as the terms may be more onerous than a negotiated investment agreement. Also, as the British example has shown time and again in the past few years, legislation can be revised and is anything but a guarantee against politically motivated changes.

Although foreign (and domestic) petroleum companies need to remain aware of political developments, their real concern is *how* the political processes and current events affect their goals and operations. This may seem to be a myopic approach to politics. In effect, however, this is the most realistic, safest, and only legitimate means by which a foreign firm can address political questions. This approach is realistic because it focuses the company directly on what affects its interests and what we earlier called specific managerial contingencies. By reducing the likelihood that the foreign oil company will become embroiled in domestic political debates and by reducing the company's political role to addressing its interests, this tactic reduces the chance that the firm will be a political casualty or will otherwise become involved in issues that are not related to corporate operations and goals. As a foreign corporation or "citizen," moreover, the TNC has no legitimate role in the domestic political life of the host beyond that of articulating and defending the firm's narrowly defined in-terests. There is a thin, ambiguously defined line between the foreign firm's legitimate expression of its interests and interference in the sovereign political life of a state. Crossing the line may carry the penalty of political repercussions far more severe than the political changes against which the firm was trying to protect itself.

In the extractive sector, it is "the nature of the beast" that there will be changes in the investment regime covering substantial projects. Change seems to be inherent in the political economy of the petroleum and minerals industries. Most firms recognize the dynamic nature of investment agreements and legislation and accept the need for modifying the host–firm relationship accordingly. (Technically, this is a two-way street: changes may prove favorable to the company as well as to the host country.) This recognition reflects the reality of political risk in the industry and does not stem from any corporate belief that such changes are desirable.

Political risks span the spectrum from minor inconveniences to total loss of assets. The effects can be direct, as in the case of a specific change in fiscal regula-tions, or indirect. Similarly, changes may be aimed at individual oil companies or at all foreign oil firms (or all foreign firms in general). Usually, political risks stem from domestic changes. In some instances, however, international or foreign events may affect the investment climate in a foreign country. The Iran-Iraq war presents risks to all oil operations in the Persian Gulf. International agreements, such as the Andean Pact, may prompt member countries to make certain changes in their investment regimes. Politics in an interdependent world, moreover, create a measure of ambiguity in determining whether a particular political risk is domestically or internationally rooted. Restrictions on the conversion of foreign exchange and remittance abroad of profits may stem from a host state's shortage of hard currency. The TNC most likely will consider this a domestic problem and will see the risk of restrictions as a wholly domestic issue. The host government, on the other hand, may claim that the foreign exchange problem is the result of a prejudicial international economy that refuses to accept local soft currencies and carefully guards the economic hegemony of the developed countries.

In the broadest sense, there are two types of changes in the host state that might affect the goals and operations of a foreign firm.[1] The most colorful and most frequently cited in the media and texts are political instability and changes in leadership. Political instability can occur at the governmental or the popular (grassroots) level. The concept is somewhat ambiguous, implying some type of political disequilibrium, unpredictability, or noninstitutionalized variant of the norm. The norm has alternately been described in terms of the firm's home state, some ideal abstract country, or the host nation's regular political patterns. Political instability, particularly when associated with violence and domestic conflict, is attention grabbing, often commanding media headlines. In most instances, however, it is not the regular state of affairs.

Changes in government may or may not be coupled with political instability. Assassinations and coups are relatively commonplace in international politics, but they are not the normal means by which power is transferred from one leader to another. The violent transfer of power is a frequent occurrence in some countries, particularly those less-developed countries (LDCs) that have no democratic traditions and are under military rule. Sudden changes in leadership can be problematic, but so can supposedly stable electorally mandated changes. The gradual and legal rise to power of socialist governments in a number of European countries—for example, the 1981 electoral victories of François Mitterrand in France and Andreas Papandreou in Greece—has presented problems to foreign firms in a number of sectors, including petroleum.

Less flamboyant than the specter of instability are political changes in the rules and regulations governing the operations and actions of foreign firms. Changes in the foreign investment regime are less likely to attract international attention and capture headlines but are more frequent and often more risky (from the perspective of foreign firms) than political instability. Policy changes addressed to all or some companies are regular occurrences in all countries, and their impact on a petroleum or gas project can be substantial.

Political risks associated with conflict or instability are the exception, while the more mundane policy changes are the rule. The results of instability may be more sudden and dramatic than regularly implemented policy changes, but instability is less common and more diffuse in its effects. Whereas a change in import-export regulations has a clear impact on an oil company, instability is an amorphous condition that may or may not hinder operations. Few conditions can be considered more unstable than civil war, yet Gulf enjoyed profitable operations in Angola in the heart of that country's civil war. Although the Luanda government is controlled by the MPLA (the Popular Movement for the Liberation of Angola), Angola remains embroiled in a civil war (combined with skirmishes with South African troops) in the southern part of the country. Despite eight years of civil war since independence, the Angolan oil industry is prosperous and growing strongly, with extensive and increasing Western participation.

In many respects, the less-dramatic events that are crucial to corporate operations and goals are more difficult to predict than explosions of violence and political instability. Seemingly subtle changes can have substantial impact on a project. Political changes are not unrelated to each other and often occur in combination. Domestic violence and instabilities, for example, make administrative and regulatory changes more likely.[2] Predicting adverse changes in the investment regime under stable political conditions is more difficult.

In instances of either instability or investment regime changes, political risks translate into two general problems for the transnational oil firms: insufficient returns on capital and/or loss of access to funds or to crude.[3] Capital or equipment losses through expropriation or destruction mean an inadequate return on investment. By stretching our concepts a bit, concern about repatriation can be expanded beyond revenues and crude to include personnel. The repatriation of personnel rarely is blocked by the host government, but civil disorder, such as terrorism and kidnapping, can present a threat to the movement and lives of corporate personnel.

Political Risk Events

Despite what we have said about the relative importance to foreign firms of instability and regulatory changes, most corporate risk analysts continue to focus on the former as the key ingredient in evaluating foreign business environments. Macro risks (see chapter 1), whether or not they involve instability, are easier to identify and predict than micro risks. In this sense, they are similar to instability and regulatory changes, respectively, although regulatory changes covering all foreign investors must be classified as macro. It is not surprising, therefore, that stability and foreign investment climate each were identified by four out of five political risk analysts (representing resource and non-natural-resource firms) as among the four main variables in assessing overseas business opportunities. The results of this 1980 survey are as follows:[4]

Variable	Percentage Identifying Variable as One of Four Most Important
Political stability	80
Foreign investment climate	80
Profit remittance and exchange controls	69
Taxation	51
Expropriation	28
Political party attitudes toward foreign investors	24
Labor strikes, unrest	21
Administrative procedures	16
Public image of firm	5

Within the categories of risks we classified in chapter 1 there are numerous examples of specific political risk events or situations. A political risk event is any outcome in the host country that is likely to have a negative impact on the success of a foreign investment venture.[5] Political risk events are those political outcomes that threaten the interests of foreign companies.

We have identified 18 political risk events that are relevant to the petroleum and natural gas industry.[6] Although not exhaustive, our list covers the major contingencies that may arise. Most of these risk events are applicable to other sectors, particularly mining. Some of the events may be relevant only to certain types of agreements; for example, price increases may be important only when the oil company is buying crude from the host under contract or as part of an overall investment agreement. All political risk events, with the exception of war, civil disorder, and ideological change, imply specific or implicit managerial contingencies. War, civil disorder, and ideological changes are more general and may prompt more-specific risk events. With these three exceptions, each political risk event, which may be a single act or a set of circumstances, can be classified as transfer, operational, administrative/statutory, ownership, or contractual, in accordance with the categories established in chapter 1.

In no particular order, the political risk events relevant to foreign investments in oil and natural gas (broadly defined to include service agreements and purchases from host governments under contract) are identified and operationally defined as follows:

1. *Foreign war.* Military conflict between the host state and another nation might interfere with a project in two ways. If the war is fought on the territory of the host or includes an aerial or ground invasion, the oil-producing facilities (which are

likely targets of war) may be destroyed. Similarly, the opposing state may impose a blockade or otherwise restrict the shipment of oil abroad. The exigencies of war, however, might induce the host government to impose changes on oil operations as part of maintaining a war economy.

2. *Civil disorder.* This covers an array of domestic instabilities, often involving violence, that range from demonstrations and riots, to sabotage and terrorism, to armed insurrection, guerilla war, and civil war. Like the problems confronting host governments involved in a foreign war, civil disorder may prompt the government to clamp additional restrictions on the oil companies. Oil facilities may be targets of the disorder. This might jeopardize facilities and personnel and force a slowdown or stoppage in production. Pipelines carrying crude and natural gas often have been favorite targets of saboteurs. If the opposition movement is successful in deposing the established order (which usually enjoys the implicit support or at least the passive acquiescence of the oil companies generating revenues for the nation's military forces and budget), the foreign firms are in particularly troubled waters.

3. *Total expropriation.* Whether by nationalization (which is aimed at a group of companies or sector of the economy) or expropriation (of a particular firm), the host government can seize the assets of a producing company, with or without compensation.

4. *Creeping expropriation.* Rather than seizing a company's assets outright, the host might choose to make gradual, continuous inroads into the company via local participation requirements or phased-in state participation. The role, freedom of operations, and profits of the foreign firm are increasingly circumscribed, while those of the host government or local investors grow.

5. *Costly fiscal changes.* After production has begun, the host government might increase the tax burden on the companies. This can be accomplished by increasing the rates of royalties, the income taxes, or other taxes or by imposing new taxes. The effects will be decreased profitability, perhaps below the level needed to justify the initial investment.

6. *Price increases.* An oil-exporting country may decide to raise the price of crude sold to the companies under contract or extracted under an investment agreement. If, for example, a firm is entitled to recover the costs of exploration or to share in output at a rate pegged to some agreed-upon price (which probably is linked to the official government selling price), an increase in price will mean that the company is entitled to less supply (or must pay more for the same amount).

7. *Domestic price control.* As a political-economic palliative, the host government may impose a below-market-price ceiling on domestic sales of products and crude. If it is required to supply the domestic market, the foreign firm may make minimal or no profits on such sales. Low prices also will stimulate domestic consumption, which will compound the problem and might mean that reduced quantities will be available for export.

8. *Production restrictions.* To conserve for future production and consumption or to implement policy goals (that is, to alleviate downward pressure on prices exerted by abundance of supply), the host may restrict the lifting of crude. This will hinder the firm's ability to realize the advantages of an economy of scale and will restrict supply availability.

9. *Export/sales restrictions.* In conjunction with limited liftings or independent of a production cutback, the host government might restrict crude exports and sales abroad. This would translate into reduced company revenues and curtailed availability of supply.

10. *Remittance restrictions.* To conserve hard currency or restrict the flow of corporate profits abroad, the host government may limit the repatriation of company earnings. This approach may also be used to encourage domestic reinvestment of profits and use of domestic product/service suppliers.

11. *Foreign exchange controls.* For reasons similar to those for remittance restrictions, the host government may clamp controls on the convertibility of local currency to hard currency. The host might require that all company assets be held in local currency, with all foreign exchange controlled and regulated by the government.

12. *Devaluation or Revaluation.* For various monetary reasons, the host government might devalue or revalue its currency. Both actions present risks: devaluation may result in exchange losses on profits and payments remitted abroad, while revaluation (a less likely occurrence) might negatively affect local earnings.

13. *Embargoes and Boycotts.* A host government may embargo sales to particular states, while a home government may boycott purchases from certain sources. Getting caught in a foreign policy imbroglio of this sort might make an involved company a policy pawn and a likely political target (for both home and host governments). This could mean a loss of a particular source or market. The company also must be wary of secondary or tertiary sanctions being imposed on it by a host or home government. Similar problems could erupt in a war situation between oil-producing states or between an oil-producing and a consuming country.

14. *Reinvestment requirements.* To compel renewed investment in oil and gas or in related sectors, the host may require that a percentage of profits or a specific sum be reinvested annually. As a related approach, the government can insist on associated investments or preinvestments as a condition for a straight sale of crude or for earning an investment contract.

15. *Domestic refining and shipping demands.* To capture some of the value added and downstream profits, the producing government might insist on exporting a higher percentage of refined products and on the use of national flagship tankers. This would result in reduced profits for the companies, as well as further cutting the operating capacity of company refineries and tanker fleets.

16. *Government-to-government sales policies.* Without reducing liftings or exports, the host government might strive to increase direct sales from its national oil com-

pany to consumer governments. This could mean a phasing out of the company from certain markets, resulting in reduced supply and earnings.

17. *Ancillary demands.* When conditions permit, the host government may press the company to make new concessions in a number of areas, including the training and hiring of nationals, compulsory subcontracting and purchasing, and the transfer of technology. The results may be increased operating costs, delays, procurement problems, or loss of technological advantage.

18. *Ideological change.* As a last, umbrella risk, which might be associated with virtually any of the preceding policy preferences, the host government may undergo a radical swing to the left and take a more anticompany stance. This may be a slow, evolutionary change ushered in by popular discontent and new leadership. An oil company is a vulnerable and likely ideological target to use as an example or a scapegoat.

Instability and Civil Disorder

Although instability is a less common problem than policy changes, it presents more far-reaching and costly risks to TNCs. Without reinforcing the overconcentration on instability, we want to elaborate on the concept and explore some related problems.

A state's political stability is its tendency to adhere to regular, institutionalized patterns of behavior. A stable system is adaptable and can resolve the problems of conflicting demands and interests through regular political and social channels. Stability implies a condition of equilibrium in which the government is not subject to the threat of violent or broad-based nonviolent opposition.

Political instability, on the other hand, is associated with frequent, large-scale, and potentially violent disturbances. Stability/instability is a continuum: no system is wholly unstable, and even the more than 200 years of continuous political tradition in the United States does not mean that the United States shows no evidence of instability. Equilibrium is a fluid concept, implying ongoing adaptation and assimilation, not a static or precarious balance. A stable state can maintain its equilibrium—the orderly processes of its institutions, institutional roles, and patterns of behavior—without abrupt dislocations or substantial challenges to its legitimacy.

To be fair, civil disorder and instability must be assessed in the context of a particular system and culture: what is disorderly and destabilizing in one society may be the norm in another.[7] Although the practices of one country cannot be criticized in terms of the values, standards, and political culture of another, defining the norm in many states is not an easy task. Can the regular use of violence and armed opposition to the government ever become the norm in the traditional sense? Though not incorrect, there is something conceptually perverse in asserting that war is the normal state of affairs between Israel and the Arab states. When the exception becomes the rule, notions of normality may become questionable.

At its most extreme, political instability might lead to "radical political change"—"the ascendancy to power of a person or group holding a different political philosophy than the person or group that it replaced."[8] This change can be electoral, but it often is wrought by violence and intrigue.

Superficial analysis has often assumed a direct, positive correlation between political instability and political risk. Although instability creates uncertainty and tension, it is not clear that instability is a direct source of risk to foreign firms. In many instances, the disorder and instability are short-lived or too insubstantial to have any effect on the operations of foreign companies. In other cases, the disorder is effectively contained by host government police or armed forces or is not directed toward foreign firms. Moreover, disorder may spring from either end of the political spectrum and can usher in a more-favorable investment regime as well as a more-hostile regime. Even in instances of radical political change, there may not be any increased risks confronting oil companies, although there may be an increased sense of uncertainty. Instability and civil disorder, however, can be the source of concrete political risks to foreign oil firms. At the local level, governments in some developing countries may not be able to enforce national policies effectively in all regions of the country. Disorder and instability may be manifested in violence that interferes with corporate operations. Oil installations and facilities may be military targets of insurgent groups. Corporate personnel may be attacked or seized by terrorist groups to obtain ransom or to exact political concessions. An Exxon line manager in Argentina, for example, was kidnapped by a local group in December 1973. The man was released in the spring, after Exxon paid a handsome ransom and sponsored advertisements advocating the position of the guerillas. Not only did Exxon have to contend with the physical risk presented by the kidnapping, but the company also had to cope with an Argentine government that was staunchly against acceding to the kidnappers' demands and threatened punitive actions against Exxon.

Oil fields are physically vulnerable to disruption. Even more vulnerable, however, are pipelines. Often spanning hundreds or thousands of miles of uninhabited or sparsely populated area, with some lines running above ground, pipelines are easy targets for guerilla groups. The Tripoli pipeline carrying Iraqi crude to northern Lebanon, for example, was bombed in January of 1982—ten days after it reopened following a five-year closure. Three months later, the pipeline carrying Iraqi oil through Syria was shut by the Syrian government, which supports Iran in its war with Iraq. Similarly, in 1981 the Guatemalan pipeline connecting the Rubelsanto oil field with the Caribbean was subjected to sabotage. In December 1979, protestors in the Indian province of Assam seized the key pipeline in the region. The protest grew, stopping all exploration and production in Assam and forcing closure of the refinery at Digboi. Order was restored only with the assistance of the army, but the disturbance lasted more than four months.

In addition to getting caught directly in the line of fire, the companies must cope with potential policy changes that may be implemented by the government in response to the instability. Depending on the extent of the disorder and whether it is violent, receives international support, and creates an economic burden on the host state, the foreign oil firms may be subject to regulatory changes designed to generate additional revenues or to restrict the outflow of earnings and hard currency.

At times, foreign companies are part of the instability. Nationalist sentiments in favor of expropriation may boil into widespread dissatisfaction. To assuage nationalist claims—or to use the company as a scapegoat—the government may expropriate a firm's assets or place it under a severely adverse regulatory regime. The oil companies tend to be so large and visible that they are likely victims of nationalist aspirations.

Operating in a country controlled by a government of questionable legitimacy and subject to extensive violent opposition presents a unique combination of risks to an oil company. If the investment climate is suitable, the firm wants to produce and sell crude, pay taxes, and earn profits. To whom the taxes are paid is irrelevant to the company. If the government depends on oil revenues, the opposition movement will most likely consider any firm that pays revenues to the government as "supporting" the hated leadership. The oil company has no choice but to pay royalties and taxes to the official government, even if its legitimacy is open to question. To do otherwise would be to court retaliation by the government for supporting the opposition.

Should the opposition be successful, the company must try to convince the new leadership that it is a good "corporate citizen" of the state and that it was not a political supporter of the old government. This can get very complicated and sensitive. If the new leadership is ushered in by violence and is ideologically opposed to the old government and the operations of the foreign oil companies, the risks will be increased accordingly. As in the case with instability in general, ideological hostility may or may not have a tangible impact on the oil firms. The socialist ideology of many revolutionary movements often becomes more rhetoric than substance when the movements become governments, with responsibility for national welfare.

Angola is again an illustrative example. When Portugal withdrew from its former colony, many analysts thought Gulf (which had producing assets in Cabinda) would be wise to withdraw from the country until the dust settled on the civil war. Three guerilla/liberation groups—the MPLA (the Popular Movement for the Liberation of Angola), the FNLA (the National Front for the Liberation of Angola), and Unita (the National Union for the Total Independence of Angola)—were vying for power. The MPLA, which enjoyed Soviet and Cuban support and extensive backing in Africa, emerged as the dominant power and installed itself in the seat of government in Luanda. Gulf resumed paying royalties and taxes as provided for by its concession agreements.* Despite its avowed socialism, the Angola government has not rejected the role played by the Western oil companies. In fact, the government has welcomed the participation of foreign oil companies in the oil and gas industry.

Modernization

A special type of instability commonly is associated with modernization. Modernization is the process of social, economic, and political development, the means by which a society's institutions and culture adapt and change in response to the demands of the contemporary world.

Although modernization is an ongoing process, the developed countries are at the apex of modernity. Developing countries, on the other hand, are striving toward development, toward modernity. Modernization requires far-reaching changes in the structure of society and its institutions. Developing states are moving—or trying to move—through a transitional stage into the modern world. As such, Third World societies, which are prone to rapid change, are seen as more unstable than developed countries.

Although it is an overgeneralization to assert that developing countries are more unstable (or less stable) than developed states (see chapter 2), the modernization process often is a period of social fragility and vulnerability to destabilizing forces. To modernize effectively, a state's leaders and institutions must be responsive and adaptive. In many instances, leaders or institutions refuse to cede their bastion of privilege and are unresponsive to popular changes. Eventually, as Iran and Nicaragua have proved and El Salvador will reaffirm, a developing state with an unresponsive leadership is torn asunder, often in an agonizing civil war. Responsive leaders, moreover, are not always able or successful at adaptation. Changes may be too fast or too slow, falling short of the demands of the society or racing ahead and challenging deep-rooted traditional values. In either event, the society may be prone to instability and civil disorder.

Another problem that seems endemic to modernizing nations is uneven development—a condition of imbalance between the levels of modernization of various parts of the economy, political institutions, and society. This is a typical

---

*The Angola case also reveals the potential cross pressures—and political risks—that oil companies can be exposed to by virtue of home government policies. The Ford administration tried to force Gulf to pay royalties and taxes to the U.S.-supported FNLA/Unita coalition (whereupon Gulf placed the money in an escrow account pending resolution of the civil war). If the United States had pursued a course of active intervention in Angola, moreover—as President Ford and Secretary of State Kissinger had recommended—Gulf would have been caught in a unique political bind that might have wrought backlash from both the Angolan and the U.S. governments.

phenomenon in Third World nations that are areas of substantial foreign investment in the extractive sectors; modern, large-scale, capital-intensive mining or petroleum industries serviced by advanced technology, extensive infrastructure, and trained labor (including nationals) grow up side by side with the traditional labor-intensive subsistence farming, plagued by shortages of modern equipment and techniques, backwardness, poverty, and poor living and commercial conditions. With enclave petroleum or minerals industries, some developing states have seen the rise of dual economies—the modern and traditional—in a tense, ironic coexistence. As oil wealth begins to spill into the economy, fueling popular expectations, introducing foreign ways, speeding inflation, and creating vast opportunities for corruption and patronage, many countries become powder kegs of instability that could be sparked at any time.

Some analysts claim that uneven development is an essential, though not sufficient, condition for extensive political risk to foreign investors.[9] According to this position, in addition to uneven development, the state must have sufficient power to challenge the TNCs. In this model, strong, unevenly developed countries are the highest-risk countries. Although uneven development is not a prerequisite for political risk, with the United Kingdom and Canada as salient examples to the oil industry, there is a great deal of validity to the contention that states must be sufficiently strong to be able to mount an effective challenge to the traditional role exercised by foreign corporations. This strength is a combination of economic and market power and social and political cohesiveness. Hopelessly poor and weak states are unable to impose demands on the companies.

This does not mean that weak, underdeveloped or unevenly developed countries are risk free; to the contrary, although governments in these countries may lack the capacity to assume control over the domestic oil industry or to exact changes in the regulatory regime governing the industry, the country may be prone to the instabilities and disorder of domestic strife and violence. As such countries become stronger and develop a national oil industry, moreover, the foreign firms may be confronted with a more-capable developing state that is both willing and able to impose changes on the operations of oil companies.

Modernization and development are processes of change. The developed countries had the luxury of time and markets to modernize in an evolutionary manner. Pressured on all sides, however, Third World states have a very difficult time keeping a steady, balanced pace of development. Often, developing countries appear to be in a condition of flux, with frequent changes of government and the use of violence. In these countries, the evolutionary model may be discarded for the revolutionary. By itself, revolution is not proof of political risks to foreign oil and gas operations and interests. Revolutionary changes often are accompanied by an anti-Western ideology and the use of force, however, and may increase the likelihood that petroleum facilities and projects will fall prey to physical threat and regulatory changes.

Policy Changes

Policy formulation and implementation are among the most fundamental responsibilities of government. Given a set of conditions, preferences, constraints, and capabilities, policies are the means by which government tries to achieve its goals. If the conditions or environment surrounding an issue, the preference or goals of decision makers, or the constraints and capabilities of the state change, this will be reflected in a revision of policy.

Policy changes are common. They may occur suddenly or be phased in over a protracted period of time. The majority of political risks we identified are regulatory in nature. Changes are inherent in the regulatory regime governing foreign investment. Even though changes may prove favorable to corporate interests, companies are concerned with the political risks: the likelihood that policy changes will adversely affect company operations.

The dynamic of negotiations and agreements in the extractive sectors is that the reality reflected in an agreement reached in the preliminary stages of project development is not the same as the reality the company must confront after production has begun. In terms of the aforementioned policy variables, once a field is online the host has fewer constraints and increased capabilities. Conditions are more favorable than they were before completion of the project, although there may be many mitigating factors, such as global supply and demand. Government preferences also are likely to undergo change; no longer anxious to consummate a deal to develop a new field, the host may concentrate on maximizing its control over and its rewards from the producing project.

Change is inevitable. As we discussed in chapter 2, the bargaining power of host and firm change during the course of a petroleum or natural gas development project. Agreements tend to be concluded under uncertainty: neither firm nor government can be certain whether a project will be successful; and if it is successful, neither can be sure of how profitable it will be. In advance of production and knowledge about market conditions, it is virtually impossible to know how to structure the regulatory regime for a particular project.

The national oil regime is especially prone to change because of the central role petroleum plays in the economy of all crude-exporting countries. This is particularly true for Third World oil producers, many of which have economies and governments that are more than 90 percent dependent on oil. Even in more-diversified, advanced economies, the petroleum industry is so valuable and important that it is a natural target for government regulation. Most governments are reluctant to permit such an essential industry to be privately controlled, particularly when the private interests are mostly foreign. Government control, therefore, is the norm; and along with government regulations and policies come changes in the regulatory regime.

The 15 political risk events related to particular managerial contingencies (we again have excluded the less-than-specific consequences of war, civil disorders, and changes in ideology) can be related to host government implementation of the policy goals identified in chapter 2 (see table 2.1) in a number of ways. Different policy goals will be related to different types of changes in the petroleum regime (political risk events), although some changes might be employed in conjunction with any of a number of national policies.

To stay in office, a government is likely to do almost anything that is necessary. Policy changes to address actual problems or satisfy popular aspirations are common responses. When petroleum and foreign interests are involved, transnational oil companies must be wary of potential risks to their operations. Although none of the political risk events are directly related to a government's objective of remaining in power, many of them may be means of strengthening the government or defusing popular opposition. Increasing the government take, for example, will give leaders a larger budget, while expropriation or domestic price controls may placate an ill-tempered populace.

Maximizing the economic returns that accrue to the host country—the main and first policy goal pursued by leaders who are not preoccupied with holding office—implies direct risks for foreign petroleum and natural gas producers. The two most obvious associated political risk events for which firms must be prepared are costly fiscal changes and/or price increases. It is unlikely that the fiscal regime for crude will ever change again as radically as it did during 1974, when OPEC country royalties climbed from 12.5 percent, to 14.5 percent, to 16.6 percent, to 20 percent, and income taxes were hiked from 55 percent, to 65 percent, to 85 percent (the cost of equity crude rose from about 60 to 88 percent of the posted price, plus production and operating costs), but costly tax and royalty changes still may occur. This is particularly true in countries that are just developing their oil resources and cannot yet command the OPEC model fiscal regime. Host countries also can impose additional fees and taxes that prove costly to corporate operations and generate revenue for the state.

Price increases are another means of boosting host state earnings. Whereas tax increases raise the *share* of the pie flowing to the host, higher prices expand the *size* of the pie. Current and near-term market conditions do not forebode any price jumps; in fact, declining real and current prices are anticipated in the short term, and spot prices, which are an important barometer of future trends, reveal a downward pressure on prices. A timely crisis could, of course, change conditions of oversupply into shortage and panic buying. Even if supply continues to outpace demand and to encourage a drop in prices, the OPEC nations and other exporters will try to hold the line against any substantial price backslide, although spot prices may plunge dramatically. The support Saudi Arabia and Kuwait lent Nigeria in 1982 to help Nigeria maintain liftings by and sales to foreign firms indicates the firmness of OPEC's desire to maintain the current price structure. As market conditions change in the mid-1980s—particularly as the world economy recovers and demand for oil grows—crude and natural gas prices may resume their price incline. It is unlikely that any future increases will be as severe as those in 1973–1974 in relative terms (when prices quintupled) or as the increase in 1979 in absolute amounts (average world prices rose by more than $13).

Expropriation or nationalization may or may not increase the earnings realized by the host government. The assumption of risks and costs by the government may offset or even the additional income realized on state-owned operations. By definition, sudden expropriation presents greater short-term risks to foreign firms. From the point of view of the host, creeping expropriation—or phased-in participation—is a more sensible approach to boosting government earnings by taking advantage of corporate expertise and capital assets. Regardless of whether or not expropriation increases host revenues, governments often *think* that it will. They can be proved right or wrong only after the fact and after the companies have absorbed the loss.

To capture more of the value added on downstream operations, the host may insist on refining crude domestically and exporting products. This approach can be pushed one step further—shipping crude or products in tankers owned by the state. To realize fully the profits from their petroleum, the host may opt for direct government-to-government (noc-to-noc) sales. This eliminates the role of middlemen, permitting the host to capture all the revenues from a sale.

Depending on the policies and goals involved, the pursuit of domestic policy goals may be linked to changes in the petroleum investment regime and can create an array of political risks. If the implementation of government initiatives requires additional revenue, the host may be inclined to pursue further profit maximization tactics. To assuage latent public unrest, to try to keep a lid on cost-of-living increases (oil exporters often suffer high rates of inflation), or to stimulate domestic energy consumption, the host may impose price controls on domestic sales of crude and/or products.

There is an implicit tradeoff between lifting and selling crude now and leaving the oil in the ground for future production. When prices are rising, oil in the ground often is a better investment than crude production; that is, the value of reserves (future production) may be appreciating faster than the income received on current sales and invested at the prevailing market rate. Not unrelated to this, production cuts can be used as a means of alleviating the pressure of oversupply or can further tighten the supply-demand balance. Although it may be too small to have the desired effect, the March 1982 OPEC decision to restrict output to under 17.5 million barrels daily is an effort to reduce the amount of excess crude on world markets and the accompanying downward pressure on prices.

Governments have other policy concerns, unrelated to pricing and investment strategies, that may justify a cutback in production. Even for countries that have reserve/output ratios extending until the middle of the twenty-first century, such as Kuwait and Saudi Arabia, conservation is an important issue. Those oil exporters that have shorter reserve/output ratios, such as Nigeria, Gabon, Venezuela, Indonesia, and Ecuador, are naturally particularly concerned with conservation. Conservation—in this instance, restricting crude liftings and leaving producible oil in the ground—is a means of stretching the petroleum-related cash flow farther into the future and of increasing the amount of crude that can be extracted from each field. Conservation also assures the local economy of sufficient future petroleum supply increasingly essential as the society develops and industrializes. Moreover, it is a way of passing on the patrimony of the land and its resources to future generations.

Exports or sales abroad may be limited, either to divert additional supply to domestic markets or, again, as a means of decreasing the amount of crude flowing onto the world market. To bolster the national economy, the host may decide to restrict the outflow of remittances or to impose more onerous controls on the use of foreign exchange. This is not an uncommon answer to the problem of hard currency shortages.

Host governments typically want greater control over their petroleum industry. Part of exercising that control—one of the principles of permanent sovereignty over natural resources—may be policies that require domestic refining and export of products rather than crude, the use of national flagship tankers, and direct sales to foreign governments or consumer nocs, bypassing the transnational oil firms. Similarly, the host may press the foreign company to promote national or regional business development by using local goods and services. To increase national oil capabilities, the state is likely to insist on the training of nationals, a typical procedure of long standing, and the transfer of technology and know-how.

An obvious means of assuming national control is expropriation or nationalization. This may or may not be accompanied by prompt, adequate, and effective compensation. In a sense, forced sale or divestiture of petroleum assets, retraction of concessionary rights, or coerced contract renegotiation may be considered expropriatory in that the firm must cede a capital or revenue-producing asset.

The pursuit of foreign policy goals also may be related to changes in the petroleum regime. Most directly, an oil-exporting country may embargo sales to certain states. As in the OPEC embargo of 1973–1974, the host governments may enlist the transnational oil companies as implementors of such policies. Oil producers also may try to use oil power as a means of influencing the behavior of foreign countries or influencing world events. This might mean changes in the operations of the oil companies or their involvement, directly or indirectly, in a foreign policy initiative of the host government.

It is in policy changes that most political risk events are rooted. From the host government's perspective, the petroleum industry is a policy tool that can be effectively used to pursue a number of goals, directly or indirectly. For the firm, this means the chronic presence of political risks.

Although our attention has been on the negative effects of policy changes, there are a number of instances when government policy initiatives have changed situations for the better, from the perspective of the companies involved. In 1982, for example, the Canadian National Energy Board announced the easing of rules restricting sales of natural gas to the United States.

A more specific example of a policy change that has had positive consequences for an oil company stems from the election of a conservative government in Norway. The labor government previously decided that Mobil was to turn over control of part of its huge Anglo-Norwegian Statfjord field to Statoil, the state-owned company, on January 1, 1985. The newly elected Storting (parliament), however, extended Mobil's control over operations until 1987. Perhaps more important, the Storting appears to be overturning the traditional policy by which Statoil was given a 50 percent stake in all Norwegian concessions.[10] These examples notwithstanding, policy changes are at the root of the majority of political risks that adversely affect company operations and interests.

Notes

1. Lars H. Thunell, *Political Risks in International Business: Investment Behavior of Multinational Corporations* (New York: Praeger, 1977), p. 7.
2. D.W. Bunn and M.M. Mustafaoglu, "Forecasting Political Risk," *Management Science* 24 (November 1978): 1562.
3. C.A. Gebelein, C.F. Pearson, and M. Silbergh, "Assessing Political Risk of Oil Investment Ventures," *Journal of Petroleum Technology*, May 1978, p. 726.
4. Stephen Blank, *Assessing the Political Environment: An Emerging Function in International Companies* (New York: Conference Board, 1980), p. 49.
5. Bunn and Mustafaoglu, "Forecasting Politcal Risk," p. 1558.
6. Our list builds on the variables identified in Gebelein, Pearson, and Silbergh, "Assessing Political Risk," and in Bunn and Mustafaoglu, "Forecasting Political Risk."
7. Stephen J. Kobrin, "When Does Political Instability Result in Increased Investment Risk?" *Columbia Journal of World Business* 13 (Fall 1978): 120.
8. Robert T. Green, "Political Structure as a Predictor of Radical Political Change," *Columbia Journal of World Business* 9 (Spring 1974): 29.
9. Howard Johnson, *Risk in Foreign Business Environments: A Framework for Thought and Management* (Cambridge, Mass.: Arthur D. Little, 1980), p. 10.
10. *Financial Times*, 24 March 1982.

III

Theory and Practice in Political Risk Analysis

6

Approaches to Political Risk Analysis

How Do We Get There from Here?

Political risk analysis is basically a two-step process: forecasting relevant political developments and evaluating the impact such developments will have on the goals and operations of a particular firm. Although transnational corporations and international business analysts are unanimous in acknowledging the importance of political variables to investments abroad and international trade, particularly in the highly politicized crude and natural gas markets, there is little agreement on how to evaluate political risks. The harmony of shared concerns about political events translates into a cacophonous array of assumptions, approaches, and models in analyzing political risks.

To some extent, the multitude of methodologies reflects the relative youth of the field; with few established techniques from which to borrow, many analysts have constructed their own approaches. As political risk analysis matures, we can expect the divergence in approaches to narrow. The kinks will be worked out of the initial efforts, which will be refined into more accurate and scientific second-generation models that build both experience and theory. The divergent strands of the field will be interwoven via the success and workability of a handful of generally recognized approaches.

In many respects, however, the different techniques also indicate the vastness and complexity of the problems under investigation. A model of political risk must operate like a crystal ball, enabling the analyst to see what a country's political climate will be at some time in the future. The farther into the future one projects, the more difficult it becomes to see clearly. To be useful, however, particularly in the extractive industries, risk analysis may need to forecast the political situation five or even ten years into the future. At the very least, a risk forecast must address the period spanning from the next six months to two years hence.

Even if the conditions are knowable, this is only half the solution. The more important question still remains: What does the changed political environment mean for the company and its projects? At this step, the political risk analyst must be prepared not only to forecast the likely problems or political risk events that will ensue but to translate events and the general political environment into specific managerial contingencies. Knowing tomorrow is not sufficient; determining how tomorrow will affect operations, profits, and other corporate concerns is the key.

The problems seem insurmountable and, contrary to our earlier statement, may prevent the field of political risk analysis from developing a coherent group of commonly accepted techniques. The nature of the problem of political risk may be irresolvable; current approaches might be supplemented by an ever-increasing sea of models trying to depict or explain the same types of phenomena. Rather than evolving into a coherent, less-diffuse business tool, political risk analysis may forever be destined to flounder on the perimeters separating the analyst from the clairvoyant, the scientist from the artist.

Despite the best efforts of practitioners to impose rigor—often mistakenly reinterpreted as to quantify—on the techniques of analysis, political risk analysis is destined to remain as much an art as a science. This by no means denigrates the field of political risk analysis or the analyst; rather, it elevates the assessment and the assessor to a station above the mechanical, requiring a mixture of insight and knowledge, perceptiveness and understanding, a sense of probability and reality. The task is complicated further by the inherently interdisciplinary roots of politics and political risks. Unlike many classroom and academic forums, there is an unclear distinction and close interdependence between questions of politics, economics, culture, and society. When this is multiplied by the divergent ideologies, leaders, types of states, economic structures, and peoples of the world, it becomes readily clear why political risk analysis requires a healthy mixture of science and art.

Assumptions and Theory

Despite repeated attempts to formalize political risk analysis in recent years, there is no general theory of political risk. By *theory*, we mean the organization and explication of the relationship between variables in a bounded area of study. An inquiry is bounded if it is distinct from that which is beyond its domain.

Theories reveal the interrelationship and connection between variables, explaining why certain results are likely to occur. Theories are explanations of reality. They are the key to organizing otherwise disparate occurrences, data, and bits of information. Being neither true nor false, the merit of a theory rests on its explanatory power: one theory is better than another only insofar as it explains more. Of course, the analyst also must take into account what it is that needs explanation.

There are no general theories of political risk. Many analysts do not bound the area of study rigorously enough to permit theory building. The dependent variable—"the vector of political contingencies arising from the political environment"[1]—often is not clearly specified. On the other hand, the dependent variable (that which is being explained) may be industry-, company-, or even project-specific, making it difficult to generalize to a broader theoretical framework. Most approaches identify the independent variables (the things that cause or lead to the dependent variable), although this also is not always the case.

There remains a major conceptual breach between identifying the field of study and the dependent and independent variables and having a theory. Just as the accumulation of facts does not yield understanding, the gathering of variables does not create a theory. As a conceptual tool to understanding, a theory organizes the variables, revealing *how* they interact and *why*. Herein lies a serious shortcoming of virtually all approaches to political risk analysis: they are conceptually unable to explain why and how a given set of political risk variables affects the interests and goals of companies.

The analysis of political risk is moving toward theory building. By this we mean that analysts increasingly are specifying the area of study and the dependent and independent variables, and, as important, they are structuring their studies. Rather than being totally eclectic and simply gathering all potentially relevant information and committing it to paper, political risk analysts are becoming more systematic.

Although they fall short of theory, systematic approaches enable and invite comparisons between countries and within the same country over time. Traditional approaches more often than not were unsystematic, precluding comparative analysis. A particular investment or project, however, must be understood in the context of and evaluated against alternative opportunities, as well as against the likelihood of adverse or positive changes in the future. This requires systematic analysis.

Systematic approaches, moreover, are more explicit in their assumptions and logic. Different techniques can be compared, critiqued, and modified. The findings and forecasts of a systematic approach can be evaluated in terms of the methodology employed, creating a type of feedback loop that enables the analyst to modify the approach according to experience.

Perhaps most important, systematic approaches usually entail a model of analysis. As a picture or simplification of reality, models make it easier to compare approaches and study the processes involved. A model or analytical framework lacks the explanatory value of a theory, but it is an important conceptual tool in ordering and organizing variables and reality. Models are an essential intermediate step in creating theories. Models may also be devised to represent a theory. Several political risk models will be compared in a later section of this chapter.

Key Variables

A multitude of independent variables may be important in analyzing political risk. Which variables are most important will depend on the model and the conditions surrounding a particular investment.

We do not propose to identify independent variables that we see as being "most" important. In the absence of a particular theory or model and project, there are no most-important and less-important variables. Nor is our list exhaustive. What we propose is to list and classify those variables that are most likely to be relevant to a firm analyzing political risks in the international oil and gas industry.

The variables are clustered under three broad headings: host variables, corporate variables, and external/international variables. Host variables have been further classified as governmental/political, economic, sociocultural, and petroleum-specific. The variables are not presented here in any meaningful order.

I. Host Political Risk Variables
   A. Governmental/political
      —dominant ideology and possible changes
      —institutional development, including the strength of the legal system, the legitimacy of the government, and the degree of bureaucratization
      —instability, including the existence of disaffected groups, governmental use of coercion and suppression, and the outbreak of violence
      —continuity and changes in leadership and in the perspectives of other major political leaders and/or parties
      —nationalism
      —domestic and foreign policies, including goals and policy changes
      —governmental corruption
   B. Economic
      —economic performance, including levels and growth in GNP per capita and inflation
      —balance-of-payments and import/export concerns
      —foreign exchange position
      —public/private sector mix
      —level of development and development plans
      —government debt
      —distribution of wealth
      —role of the foreign oil firm in the domestic economy

—integration between petroleum/gas industries and the remainder of the national economy

—importance of petroleum to government revenues and the overall economy

C. Sociocultural

—homogeneity: ethnic, linguistic, racial, and national

—standard of living

—receptiveness to foreign influences

D. Petroleum-specific

—ownership

—domestic reserves/production

—host's relative market position

—level and destination of exports

—strength of the national oil company

—role of the foreign firm in the national oil industry

—prices

—domestic ability to operate the industry, including commanding the necessary skills, technology, know-how, and capital

—ownership/contractual relationship between the firm and the host

II. Corporate Political Risk Variables

—nationality of the company

—position in the world industry, including sources of crude, reserves, production, and market outlets

—special bargaining advantages: technology, managerial skills, services, and capital

—dealings with host government: receptive, diplomatic, and open, or unreceptive, brusque, and unyielding

III. External/International Political Risk Variables

—host government participation in international treaties, conventions, and organizations

—political/economic relationship between the host and the home government, including security, trade, and aid issues

—involvement of the host in international conflicts and potential results of conflicts not involving the host directly

—world petroleum market conditions: prices, supply, and demand

—world economic conditions, including economic growth and energy consumption

—developments in other oil-exporting countries (demonstration effect)

The foregoing list merely points in the direction of which variables may be important. It is the role of a model or theory to refine the list and organize the variables—their relationships, flows, and consequences—into a useful analytical tool. Simply gathering information without a framework of analysis is a dead-end path. There are too many variables, moreover, to be plugged directly into a framework. Depending on the model, some variables will be discarded or treated implicitly; others will be grouped together.

Rather than designing a theory or a broader model of political risk, some analysts have generated hypotheses directly from one or more independent variables. A hypothesis is an assertion of a relationship between two or more variables that can be verified or that is subject to proof. Drawing on empirical evidence or logical assumptions, a number of hypotheses have been advanced with respect to political risks.

Hypotheses are testable. Although there is much to be learned from hypothesizing about the relationship between an independent and a dependent variable, no accumulation of hypotheses will produce a theory. (A hypothesis also may be inferred from a theory.) Once again, there is a conceptual leap between stating a hypothesis about the relationship between two or more variables and developing a general explanation about the interaction between variables.

Perhaps the crudest (no pun intended) hypothesis about political risk is the one that claims that less-developed countries present greater risks than more-developed states. A more common variant on this asserts that traditional, premodern states and developed, consumer-oriented countries are low risks, while industrializing states are the most risky. The logic here, argued by Howard Johnson and others, is that the industrializing countries are the most prey to uneven development, which many analysts see as an essential (but not sufficient) precondition for political risk.[2] Figure 6.1 depicts the hypothesized relationship between uneven development and political risk.

Yet another variant on the hypotheses relating political risk to the type of state and level of modernization, which has been advanced by Robert T. Green, employs a seven-tier classification.[3] Green claims that the likelihood of the "risk of radical political change," defined as the rise to power of leaders with a political perspective different from that of the leaders they replaced, is related to the country's governmental form. The relationship is shown schematically in figure 6.2.

In addition to the simplicity of identifying levels of risk with different governmental forms, this hypothesis suffers from a poorly conceptualized dependent variable. As defined, radical political change may prove beneficial to foreign firms, have no impact on them, or have detrimental consequences. There is no effort to relate the possibility of change to a direct risk to foreign companies. The dependent variable also steers companies in the wrong direction, focusing them on one of the less likely (albeit more dramatic) problems with which they might be confronted.

A more empirical approach to formulating hypotheses has been used by David A. Jodice.[4] Based on data for 1968–1976, the Jodice study falls short of its self-proclaimed task—to present a theory explaining nationalization and expropriation in the natural resources sector—but does generate a series of five hypotheses. The

Figure 6.1   Relationship Between Uneven Development
and Political Risk (Hypothesized)

Level of Economic Development	Uneven Development[a]	Probability of Adverse Changes or Losses Induced by Political Risk Events
Traditional (premodern)	LOW	LOW
Take-off (industrialized)	HIGH	HIGH
Mass consumption developed	LOW	LOW

[a] Based on societal and technical achievements, resource abundance, political development, and domestic peace.

SOURCE: Adapted from Howard Johnson, *Risk in Foreign Business Environments: A Framework for Thought and Management* (Cambridge, Mass.: Arthur D. Little, 1980).

dependent variable, though only one type of risk, is clearly defined. The hypotheses are interrelated, yielding more insight than would a single variable or a number of unrelated hypotheses. Perhaps most important, Jodice tests his hypotheses against the data. The five hypotheses are as follows:

1. *There is a curvilinear relationship between modernity of host country and expropriation of foreign direct investment in the natural resources sector.*

2. *There is a positive linear relationship between state capacity (central government revenues/GNP) and propensity to expropriation.*

3. *There is a positive and linear relationship between economic performance failure (instability of export earnings) and expropriation.*

4. *There is a negative and linear relationship between government dependence on bilateral foreign aid from the United States and expropriation of American corporate assets.*

5. *Controlling for the coercive disposition of the state, there is a linear positive relationship between increases in the magnitude of political threat against governing and their propensity for expropriation.*[5]

It is interesting that hypothesis 1, which is similar to that presented by Howard Johnson, is not supported by the data. To the contrary, Jodice concluded that there is a positive and linear relationship: ''The propensity for nationalization increases with

Figure 6.2   Relationship Between Government Form (type of state and level of modernization) and the Probability of Radical Political Change (hypothesized)

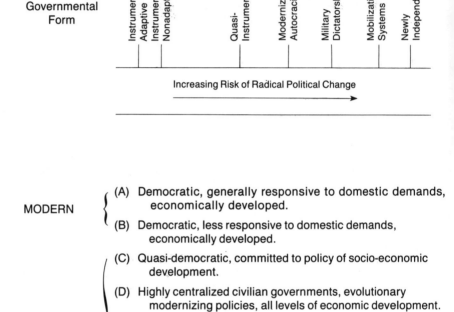

Governmental Form

Increasing Risk of Radical Political Change

MODERN
(A) Democratic, generally responsive to domestic demands, economically developed.
(B) Democratic, less responsive to domestic demands, economically developed.

MODERNIZING
(C) Quasi-democratic, committed to policy of socio-economic development.
(D) Highly centralized civilian governments, evolutionary modernizing policies, all levels of economic development.
(E) Highly centralized military governments, all levels of economic development.
(F) Extremely militaristic, headed by charismatic leader, highly ideological.
(G) Independent within last two decades, subject to changes in governmental system.

SOURCE: Reprinted by permission from Robert T. Green, "Political Structure as a Predictor of Radical Political Change," *Columbia Journal of World Business* 9 (Spring 1974). Copyright 1974 Columbia Journal of World Business.

the economic development and administrative capacity of the host country"[6]
Hypothesis 4, which also is part of the larger Johnson model (see the next section), is
only weakly supported and may be more aptly described as a curvilinear relationship.
Hypotheses 2, 3, and 5 were the most strongly supported. Although foreign firms
cannot readily employ the Jodice hypotheses as a complete model of risk analysis, the
hypotheses do point toward salient questions, and they are derived from historical
evidence.

Models

Models organize the variables being studied, usually indicating the flow of processes.
They do not include any explanation of the hows and whys, but they reveal the basic
linkage between independent (and mediating) variables and the dependent variable.
The degree of detail varies widely between models: some are very general, employing broad aggregate categories, while others use particular variables that later can
be plugged into an accompanying procedure for risk evaluation. The more systematic approaches tend to relate their model of political risks to the methodology of
evaluation.

A very general model, developed by Franklin R. Root, is presented in figure
6.3.[7] The independent variables are aggregated and are not amenable to immediate
use in any analytical procedures. Root states that the government may be an "agent
of change" or independent variable in the model, but that government behavior is
more accurately conceptualizd as a response to economic and social changes. In turn,
"government response" is mediated by its leaders, ideology, and capabilities.[8]
Governmental outputs then affect foreign firms, potentially leading to political risks,
which Root divides into three categories, similar to the classification we presented in
chapter 1.

A more detailed model of the causes of political risk has been constructed by
Howard Johnson.[9] Johnson's model (see figure 6.4) identifies five independent
variables, all of which are related to the concept and problems of nation building and
development of a modern state.[10] These five variables—the degree of national integration and common bonds, general acceptance of the government, popular con-
formance to the government's authority, participation and influence of individuals in
government decision making, and popular acceptance of the existing distribution of
material and political resources[11]—form a complicated network that determines the
pace of the country's development. Johnson measures overall national development
in terms of political, social, and technical development, resource abundance, and
domestic peace to determine the host's level of uneven development. As the model
indicates, uneven development is mediated by aid and power—the receipt or absence
of military, economic, or diplomatic support from the United States and a composite
of the host's relative economic, military, and diplomatic capacity.[12] Aid to the host

Figure 6.3   Root's Model of Political Risk

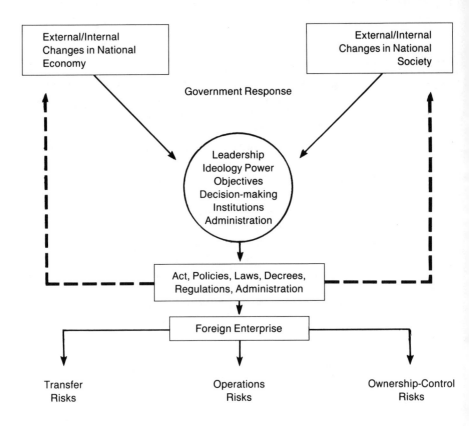

SOURCE: Reprinted by permission from Franklin R. Root, "Analyzing Political Risks in International Business," *The Multinational Enterprise in Transition*, ed. A. Kapoor and Phillip D. Grub (Princeton: Darwin Press, 1972). Copyright 1972 The Darwin Press.

Figure 6.4 Johnson's Model of Political Risk

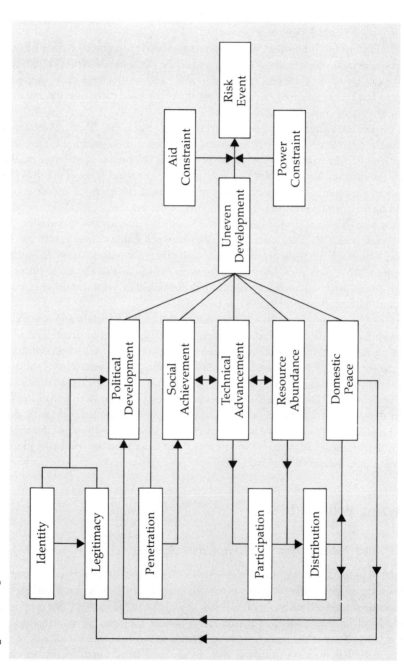

SOURCE: Howard Johnson, *Risk in Foreign Business Environments: A Framework for Thought and Management* (Cambridge, Mass.: Arthur D. Little, 1980).

country will inhibit the occurrence of "risk events" (not specifically defined by Johnson), as will insufficient host power.

The logic of the Johnson model is far more explicit than that of Root's scheme. The flow of relationships is more specific and more variables are detailed, although the components of development (political, social, and so on) tend to be aggregate concepts. The model provides one way to see the origins and causes of risk in a more than rudimentary manner.

The political risk technique developed by D.W. Bunn and M.M. Mustafaoglu (specifically designed for oil firms investing abroad) employs an implicit model of political risk.[13] Although the authors do not present a specific model per se, they identify the independent variables that may give rise to several types of political risks (dependent variables). The occurrences of ten "political risk events," outcomes that "would have a negative impact on the success of the venture," are determined by "political risk factors," those "circumstances which influence the occurrence of a political risk event."[14] The interrelationships between political risk factors are not specified, although we can presume that such interrelationships exist because of overlapping between factors. Moreover, the authors indicate that the list includes only some of the major influencing political risk factors. The prose model devised by Bunn and Mustafaoglu is presented in table 6.1.

More than in the other models reviewed, Bunn and Mustafaoglu specify the dependent variables in a way that has a direct meaning for foreign firms. The single vectors we have implied (political risk factors → political risk events) do not do full justice to the authors' framework. In addition to missing the interrelationships between risk factors, it also fails to take into account the cross-impact probabilities between risk events.[15] The occurrence of certain events, particularly civil disorder and war, increases the probability of other political risk events. This implies a rather complex relationship that we do not account for in our brief presentation of the model. Moreover, the model is designed to structure risk evaluation as well as to identify dependent and independent variables.

Analyzing Political Risk

Moving from Political Risk to Political Risk Analysis

A model of political risk is not a model of political risk analysis. A method of analysis implies some model of the processes involved, even if the model is not formalized or explicitly presented. Models of political risk are useful in themselves, but they must be translated into an analytical methodology before they can be used to analyze political risks. None of the models in the preceding section indicate the analytical procedures to be followed in evaluating risk. Even the detailed variables identified by

Bunn and Mustafaoglu are nothing but an interesting list until they are structured into an analytical methodology.

The methodology tells us *how* to analyze political risks. It is the key to structuring our thoughts, directing us toward important information, and organizing our procedures. The finished analysis is a product of the methodology. In a sense, methodology is our quality control: it ensures similarity in procedures and comparability of results.

Attempting to analyze political risks without a methodological approach is like trying to plot the path of the planets without knowing the basic principles of physics and astronomy; studying and accumulating all the facts and bits of information about the universe is the least efficient means and probably would not even tell you what you want to know. Approaching political risk analysis as an information-gathering problem is an exercise in futility. One could never obtain *all* the potentially relevant information. The costs of such an approach would be prohibitive. Even if all the information was assembled, you still would not be any closer to analyzing political risks than to predicting the course of the planets.

The traditional approach to political risk has been unstructured. Rather than analysis, the product tended to be descriptive prose—talking about many things, analyzing nothing. Some of the descriptive surveys were well researched and presented, covered the important variables, and included sensitive personal insights. Sequential presentation of large amounts of information and data, however—even if it is the necessary information—is an inconclusive procedure. Although one may intuitively derive conclusions from reading a descriptive survey (often in combination with a trip to the prospective host country), one cannot draw analytical conclusions.

In some instances, the authors of such surveys may draw their own conclusions. This is no less an intuitive approach than permitting the reader to draw the bottom line. Intuition is often an invaluable tool in evaluating political risk, but it should be used in conjunction with a rigorous analytical procedure. Relying on intuition alone is like trying to guess the position of Venus three years from today.

Descriptive summaries rarely are used as the sole criteria for evaluating political risk and the prospects for an investment abroad. In most instances descriptive summaries are prepared to accompany the more formal and structured analyses. The description usually will include important background information and will place the analysis in a country-specific context. Virtually all of the firms that sell political risk analysis services include such descriptive surveys. Ironically, descriptive surveys tend to be relied on most heavily by the transnational corporations, including oil firms, that evaluate risk internally. Many of these companies have no standard analytical procedures to assess political risk and thus produce a virtual smorgasbord of unstructured, descriptive/intuitive presentations.

Moving beyond the descriptive survey, there is a broad range of more analytical methodologies for evaluating political risk. The approaches range from relatively

Table 6.1   Bunn and Mustafaoglu's Prose Model of Political Risk

Political Risk Event ◀───	Political Risk Factors
1. Civil disorder	1.1 Strength of economy
	1.2 Aspiration levels
	1.3 Continuity of leadership
	1.4 Socioeconomic suppression
	1.5 Political suppression
	1.6 National coherence
	1.7 Regime legitimacy
	1.8 Government corruption
	1.9 External support of resistance or liberation movements
	1.10 Hostility toward home country
	1.11 Visibility of foreigners
2. War	2.1 Ideological shift
	2.2 Threat negation
	2.3 Arms race
	2.4 Negative sanctions
3. Sudden expropriation	3.1 Ideological change
	3.2 Visibility of foreigners in economy
	3.3 Colonial identification of firm
	3.4 Demonstration effect
	3.5 Marketing ability
	3.6 Bureaucratic development
	3.7 Availability of managerial and scientific technology
	3.8 Relations with firm's parent country
	3.9 Reciprocal dependence
	3.10 Performance of economy
4. Creeping expropriation	4.10 Same as for items in number 4, above, but occurring over a longer period of time

Table 6.1   *(continued)*

Political Risk Event ◄──────	Political Risk Factors	
5. Domestic price controls	5.1	Domestic inflation
	5.2	Political support
6. Adverse tax changes	6.1	Ideological shift
	6.2	Nonindustry tax revenues
	6.3	Defense expenditures
	6.4	Demonstration effect
7. Production restrictions	7.1	Large income effects
	7.2	Investment absorption capacity
	7.3	Conservation
	7.4	Production regulation
8. Production export restrictions	8.1	Internal consumption
	8.2	Participation in an embargo
9. Repatriation limitations	9.1	Balance of payments
	9.2	Economic development
10. Devaluation risk	10.1	Balance of payments
	10.2	Decline in reserve position
	10.3	Internal inflation
	10.4	Government policies that treat symptoms rather than causes
	10.5	Economic-political policies
	10.6	Economic ties

SOURCE: Reprinted by permission of D.W. Bunn and M.M. Mustafaoglu, "Forecasting Political Risk," *Management Science*, Volume 24, Number 15, November, 1978. Copyright 1978 The Institute of Management Sciences.

simple to complex, from qualitative to quantitative, from industry- or company-specific to general. The particular methodology stems from the model of political risk involved, the extent of detail and specificity in organizing the independent and dependent variables, and the preference of the analyst.

Most approaches combine qualitative and quantitative techniques. Usually, country experts are called upon to provide the data. Although these country experts may be used as information sources and may be called upon to write a descriptive summary, the data are processed by the analytical methodology that is used to evaluate risk. A panel of experts, for example, may be asked a series of questions about a country, which they are instructed to answer according to a five-point scale: definitely, probably, an even chance, probably not, and definitely not. The answers are then inserted into the analytical framework, which processes the data, standard-izes the procedures, and produces a result—usually expressed as a percentage or as some measure of likelihood of a particular event or class of events.

Problems in Analysis

Before we describe a series of techniques for analyzing political risks, a number of problems must be addressed. Building on the preceding section, the conceptualiza-tion of risk, its causes, and how to measure its causes provide the blueprint for analysis. We do not, however, have a perfect knowledge of the relationship between the political environment and the goals and interests of the firm. One can study the incidence of political risk retrospectively. Although empirical analysis of past oc-curences may yield a better understanding of the processes involved, the past is as amenable to multiple interpretations as is the future.

Moreover, the assumption that the future will resemble the past is a dangerous trap into which many analysts fall. The use of events data—the classification and ac-counting of events—and extrapolation of past events into future trends is a common practice. Although events data may reveal many things about countries and may provide a shorthand tool for comparison, there is little validity in using such data to predict the future. There is something analytically dubious about forecasting future conditions by focusing on the past.

There also is a tendency on the part of many analysts to overemphasize dramatic events, particularly incidents of expropriation and domestic instability. This results in a skewed analysis and misses the importance of policy changes and other developments. As with relying on events data, too much focus on dramatic issues and occurrences results in an overreliance on current events and the problem of predicting the future from the recent past.

Political and economic developments must be interpreted in the context of the country in which they occur (hence the importance of some qualitative, descriptive background). The stability or instability of a developing country, for example, cannot be assessed in terms of Western standards. Moreover, the real concerns of foreign firms are the implications for them. What might be important to an oil company operating in a developed country may not be relevant in the Third World, and vice versa.

While extolling the importance of well-conceptualized, clearly defined models of political risk and analytical procedures for evaluating risk, we must admit a shortcoming in our understanding of the processes involved. This is an irresolvable dilemma. Our analysis of political risk can be no better than our understanding of the problems. The methodologies employed always must be subject to improvements and change as our understanding of the phenomenon grows. Since we are not clairvoyant, we should use the model and framework for analysis that proves to be the most useful, given the circumstances.

As with models of political risk, the bottom line of a risk analysis needs to be stated in a manner that is precise and has meaning for the firm. Conclusions that in country X there is a Y percent chance of a major expropriation in year Z or that a particular country has a risk rating of 57 on a scale of 100 has little meaning for someone who is responsible for making investment decisions. Too often, a poorly conceptualized dependent variable is incorporated into the methodology. The results are ambiguous conclusions that are of little or no use.

The magnitude of risk is of obvious importance. Less often appreciated, however, is the importance of the *type* of risk. Depending on the particulars, domestic price controls or trade sanctions against a minor importer probably will be less costly to an oil-producing firm than would restrictions on liftings, exports, or the repatriation of profits. Failure to specify and distinguish between types of risks indicates a poor conceptualization of the dependent variable and results in ambiguous conclusions. More dangerous, however, is the chance that failure to identify clearly the type of risk will lead decision makers to draw incorrect conclusions. Decision makers are concerned with specific managerial contingencies in a particular context. Basing specific decisions on ambiguous, general indicators of political risk is a prescription for error and miscalculation.

Information usually is not a problem; rather, it is the organizing and processing of the information that is central. At times, sufficient information may not be available or there may be conflicting information. Getting the most current information often is a problem. Depending on what one wants to know, there are times when the information does not exist, particularly in developing countries. In other instances, the problem is in the information gathering. Some sources are of questionable validity because they have an interest in either a favorable or unfavorable evaluation, have an ideological bias, or simply are unreliable. Some governments are highly secretive and are very stingy about giving information. At times, information is skewed for political purposes. Also, of course, the same "facts" can be seen in any number of ways. Events data sets covering the same countries and time frame, for example, often show significant discrepancies because of differences between sources and researchers.

There is a tendency in risk analysis to treat what are really subjective estimates as objective data. Expert-generated data are not arbitrary or strictly subjective. However, expert opinions are not objective facts either. Most analytical techniques attempt to control for this by building a level of confidence rating into the methodology. Those projections or estimations about which the expert feels uncertain or which are most subjective can be downgraded by a confidence factor. Uncertainty should be treated for what it is and not considered a probability of any kind.

As in other areas of the social sciences, political risk analysis must cope with measurement problems. Growth in GDP or GNP, for example, is a measurable quantity, as are inflation and the amount of foreign exchange in the national treasury (although there may be information problems). But how does one measure stability or ideology? Invariably, the analyst is forced to rely on indicators—measurable variables that are assumed to indicate that which really is of interest. Instability, for example, may be measured or operationalized in terms of the frequency of civil violence, the number of people involved, and the number of deaths. Ideology, on the other hand, may be measured by means of content analysis of the speeches made by government leaders—counting the frequency and intensity of the invoking of ideological principles.

There always is a gap between the indicator and the variable being measured indirectly. It is inevitable that questions will arise about whether the analysts *really* are measuring what they claim or are using a poor indicator. Even the best indicator falls short of the underlying variable. Direct measurement of many variables is impossible, and the analysts must rely on imperfect indicators. The only good approach is for analysts (and users of the analyses) to be aware of the problem, select indicators carefully, and recognize the likely breach between the indicator and the variable that they are trying to gauge.

Quantitative approaches introduce a set of solutions to oil problems and a bundle of new concerns. Borrowing from econometric and statistical procedures, as well as from "politimetrics"—the quantitative study of political entities and processes[16]—there has been an increasing tendency to apply quantitative approaches to the analysis of political risk. Quantitative techniques reduce the interference of subjective evaluations, adhere to a proved and acceptable procedural logic, maximize the standardization of methodologies and comparability of conclusions, and are more easily understood by business decision makers and used in the decision-making process.

Quantitative methodologies, on the other hand, may only be a veil for subjective evaluations. Numbers usually are treated as hard, objective data. In practice, however, most quantitative techniques implicitly reflect a subjective model or procedure. The quantification of opinions can be no more objective than the opinions it represents. All too often, quantitative methodologies are presented *ipso facto* as more rigorous, objective, and analytical than nonnumerical approaches. Numbers sometimes are used to hide uncertainty and misleadingly to translate this uncertainty into a specific measure or probability. One analyst has gone so far as to conclude that the use of quantitative procedures to forecast political risk "is nothing else but systematized subject judgment about the future."[17] Quantitative methodologies also are prey to the data and measurement problems discussed earlier.

Overall, the biggest problem to be overcome is conceptual: How should the independent variables be organized and measured and the dependent variables operationalized? This gets back to the problem of lacking a general theory of political risk to direct political risk analysis.

Methods of Analysis: A Survey

Having discussed models of political risk and keys and problems in designing analytical techniques, we now must address the science or art of putting it all together. The methods often are classified into groups according to distinguishing characteristics. Stephen J. Kobrin, for example, classifies methodologies according to their data source (observational/expert), their breadth and depth of coverage (in-depth, selected countries/cross-national scan), and whether they are structured/unstructured (specify cause and effect and underlying assumptions) and systematic/unsystematic (employ formal forecasting).[18] Other classifications, such as qualitative/quantitative, are more simple, but may be less useful analytically and may create gross categories of disparate approaches.

Rather than classify, we have ordered the analytical methodologies on an intuitive scale ranging from those procedures that are the most general in application to those that are most specific. To a large extent, our scale incorporates Kobrin's variables, but less rigidly and less formally. We are more concerned with the specific conceptualization of the analytical technique as a means of evaluating political risks in a way that is meaningful for firms operating abroad.

The simplest ways to evaluate political risks have been referred to as the "grand tours" or "old hands" approaches.[19] The former relies on a visit to the prospective host country by corporate decision makers and the latter on expert opinions; neither approach employs a specific analytical procedure. The results are highly dependent on the individuals involved and are not comparative across countries or over time. These are the most-subjective, least-systematic ways to approach the problem of analyzing political risk. These approaches can best be characterized as unstructured, descriptive, intuitive, and subjective.

The methodologies that follow use some type of expert opinion or modified Delphi technique and/or quantitative methodology. The most basic expert opinion approach is simply to ask an expert or a series of experts to answer particular questions about a country, often in the form of a checklist. The results are then combined into a single score of some kind. The Delphi method is more advanced. Specific issues or events are identified, and the experts rank the variables according to those that are the most likely to occur and the most important to foreign firms. The opinions are aggregated into a single conclusion or range of probabilities. A level of confidence or a feedback procedure often is used to upgrade the quality of the analysis.

Another general approach to assessing political risk is simply to rate countries according to the results of a series of questions and answers. The result is described as a political risk rating. Usually, the higher the scaled rating, the lower the political risk is supposed to be. Table 6.2 presents a sample rating scale for classifying a country according to political risk.

The two most popular commercially available political risk ratings—the Business Environmental Risk Index (BERI) and the Business International Country Ratings (BI)—use a similar type of rating scale. BI uses a "scoreboard approach" of three subindices comprising ten factors each and a fourth subindex of four factors.[20] The first three subindices are based on political, legal, and social; economic; and financial factors; the remaining index (added to the rating in 1979) is energy vulnerability (dependence on imports of oil). The three major subindices have a maximum value of 100 points, and the fourth subindex is worth a maximum of 40. "Departure from the perfect score (340) reveals the degree of risk and lessened opportunity—political, economic and financial—that foreign investors face in a specific country.[21] Each factor has been assigned a point rating, with fixed values for the "most favorable" (the maximum), the "most undesirable" (the minimum), and intermediate situations. The combined score of the four subindices is averaged (by dividing by 3.4) to yield a country rating based on a scale of 100. Table 6.3 shows the factors and weights attributed to each of the subindices.

The BERI index is based on 15 variables, each of which is rated on a scale from zero (unacceptable conditions) to 4 (superior conditions). A perfect score would be 100. The variables are clustered into political, operations, and financial environmental risk subindices.[22] A permanent panel of 105 volunteers is employed to ensure continuity.

Although the BERI methodology has no explicit model of political risk or how such risk might affect a foreign firm,[23] F.T. Haner, the president of BERI, Ltd., subsequently outlined a revised framework that uses a more formal model.[24] The model (see figure 6.5), however, is conceptually confusing. The independent variables are separated into "symptoms" and "causes," with no accompanying explanation or justification for including or excluding variables. The dependent variable is left unspecified, except as an undifferentiated notion of political risk.

The Haner framework uses a two-step process and projections for three different time periods. Each of the ten variables is rated from zero through 7 for each period. Unlike the other ranking techniques, a lower number indicates a more favorable climate or reduced political risk, while higher numbers mean higher risks. An additional 30 points are assigned, at the discretion of the analyst, "to causes that should be given additional weight because of their overwhelming impact on overall political risk.''[25] The result is a 100-point scale for each time frame.

All four country political risk rating indices suffer from similar problems. All fail to conceptualize the dependent variable explicitly and make no effort to relate political risk to any specific contingencies that may confront foreign firms. The gross scales have no project-, company-, industry-, country-, or risk-specific applications and cannot easily be used in decision making. The models of political risk either are treated only implicitly or are weak in detail and/or conceptualization.

More poignant, however, are the problems associated with the scoring. The points assigned to various factors, variables, and subindices seem arbitrary and highly subjective. Why some factors are weighted more heavily than others is not explained or justified. Within each factor, the values assigned to the different answers also appear arbitrary. The scales give no indication of measurable magnitudes. The distances between integers have no meaning other than more or less risk and are not comparable.

The terms employed, furthermore, are not sufficiently defined and mean different things to different analysts. Masking this lack of definition with numbers glosses over differences in understanding, with the illusion of hard numbers that are objective indicators. Overall, the rating indices use poorly conceptualized models that are subjectively and even capriciously measured.

Political risk rating scales have some usefulness in identifying and tracking variables that are important. Changes in ratings may be important indicators of potential trouble ahead. The methodology can be adjusted to reflect more accurately the concerns of individual firms; for example, additional variables can be added (or others subtracted) to better approximate the particular interests of a company that is involved in an investment project abroad. Similarly, the firm can change the weights and scoring of variables to emphasize (or deemphasize) those issues it considers more (or less) important to its operations abroad.

Building on the groundwork laid by the concept of a rating scale, Derek F. Channon and Michael Jallard have suggested that political risks should be evaluated in terms of the goals and interests of the host government.[26] Their "political acceptability profile" is based on the assumption that "individual nation states will regard particular investments with more or less favour dependent upon their own goals."[27] State goals with respect to foreign investment are operationalized in ten factors that are clustered into three categories: economic, national sovereignty, and ideology/culture/foreign relations. These variables can be plotted for a given host country and compared to the characteristics of a particular project. Figure 6.6 presents Channon and Jallard's political acceptability profile for three hypothetical types of countries.

There is something immediately attractive about being able to see an assessment graphically rather than mentally comparing rows of numbers. The political acceptability profile is designed to be tailored to the prospective host country and the par-

Table 6.2 Corporate Rating Scale for Determining a Country's Investment Climate

Item	Number of Points — Individual Subcategory	Range of Category	Item	Number of Points — Individual Subcategory	Range of Category
Capital repatriation:			**Political stability:**		
No restrictions	12	0–12	Stable, long term	12	2–12
Restrictions based only on time	8		Stable, but dependent on key person	10	
Restrictions on capital	6		Internal factions, but government in control	8	
Restrictions on capital and income	4		Strong external and/or internal pressures that affect policies	4	
Heavy restrictions	2		Possibility of coup (external and internal) or other radical change	2	
No repatriation possible	0				
Foreign ownership allowed:					
100% allowed and welcomed	12	0–12	**Willingness to grant tarrif protection:**		
100% allowed, not welcomed	10				
Majority allowed	8		Extensive protection granted	8	2–8
50% maximum	6		Considerable protection granted, especially to new major industries	6	
Minority only	4				
Less than 30%	2		Some protection granted, mainly to new industries	4	
No foreign ownership allowed	0		Little or no protection granted	2	

Table 6.2 *(continued)*

Item	Number of Points Individual Subcategory	Range of Category	Item	Number of Points Individual Subcategory	Range of Category
Discrimination and controls, foreign versus domestic business:			Availability of local capital:		
Foreign treated same as local	12	0–12	Developed capital market; open stock exchange	10	0–10
Minor restrictions on foreigners, no controls	10		Some local capital available; speculative stock market	8	
No restrictions on foreigners, some controls	8		Limited capital market; some outside funds (IBRD, AID) available	6	
Restrictions and controls on foreigners	6		Capital scarce, short term	4	
Some restrictions and heavy controls on foreigners	4		Rigid controls over capital	2	
Severe restrictions and controls on foreigners	2		Active capital flight unchecked	0	
Foreigners not allowed to invest	0		Annual inflation for last five years:		
Currency stability:			Less than 1%	14	2–14
			1%-3%	12	
Freely convertible	20	4–20	3%-7%	10	
Less than 10% open/black market differential	18		7%-10%	8	
10% to 40% open/black market differential	14		10%-15%	6	
40% to 100% open/black market differential	8		15%-35%	4	
Over 100% open/black market differential	4		Over 35%	2	
			Total		8–100

SOURCE: Robert B. Stobaugh, Jr., "How to Analyze Foreign Investment Climates," *Harvard Business Review* 47 (September–October 1969): 100–108.

Table 6.3 Bi Country Ratings: Factors Rated and Maximum Scores

Political-Legal-Social Factors	Score
1) Political Stability	
a. Long-term stability guaranteed	15
b. Strong government but vulnerable institutions	10
c. Elected government prone to ideological swings	8
d. Active internal factions	5
e. Strong probability of overthrow (external and internal)	2
2) Probability of Nationalization	
a. No threat	15
b. State participation in selected firms	12
c. Full takeover of specific firms	9
d. Nationalization of key sectors	6
e. Large-scale nationalization	3
3) Restrictions on Capital Movements	
a. No restrictions on any transfers	15
b. Minimum controls	12
c. Limits on certain inflows and/or outflows	9
d. Strict restrictions on remittances and repatriation	6
e. No transfers permitted	3
4) Desire for Foreign Investment	
a. No restrictions on any type of foreign investment	10
b. Favorable climate with incentives	8
c. Selective investment policy	6
d. Lukewarm climate for foreign capital	4
e. Hostile foreign investment climate	2
5) Limits on Foreign Ownership	
a. No ceiling on foreign equity percentage	10
b. Desire but no requirement for local equity	8
c. Local majority required in many or key industries	6
d. Strict joint venture requirements	4
e. Only foreign minority position tolerated	2
6) Limits on Expansion of Foreign-Owned Firms	
a. No government restrictions on expansion	8
b. Obstacles to expansion, e.g. antitrust, environment	5
c. Limits on specific industrial sectors	3
7) Government Intervention in Business	
a. Free enterprise system	8
b. Limited government controls, e.g. price controls	6
c. Strong but selective government intervention	4
d. Tightly controlled economy	2
8) Likelihood of Internal Disorder and Vandalism	
a. No threat of disorder	8
b. Isolated cases of unrest	6
c. Strong possibility of vandalism, kidnappings	4
d. Probability of social revolution or civil war	2

Table 6.3 *(continued)*

Political-Legal-Social Factors(cont'd)	Score
9) Delays in Getting Approval	
a. No red tape	6
b. Occasional delays	4
c. Exasperating red tape	2
10) Cultural Interaction	
a. Easy to grasp cultural and business concepts	5
b. Difficult to establish confident rapport	3
c. Impossible to assimilate country's culture	1
Commercial Factors	
1) Present Market Size as Indicated by GNP	
a. Above $500 billion	12
b. $100-$500 billion	9
c. $50-$100 billion	6
d. $10-$50 billion	3
e. Less than $10 billion	1
2) Annual Average Real GNP Growth—Last Five Years	
a. High growth (above 8%)	6
b. Good (5-8%)	4
c. Moderate (3-5%)	2
d. Low (less than 3%)	1
3) Annual Average Real GNP Growth—Next Five Years	
a. High growth (above 8%)	8
b. Good (5-8%)	6
c. Moderate (3-5%)	4
d. Low (less than 3%)	2
4) Present Market Sophistication as Indicated by Income per Capita	
a. Above $6,000	12
b. $4,000-$6,000	9
c. $2,000-$4,000	6
d. $1,000-$2,000	3
e. Less than $1,000	1
5) Restrictions on Foreign Trade—Next Three Years	
a. No restrictions	12
b. Minor restrictions	
c. Substantial restrictions	
d. Stiff restrictions	
6) Availability of Local Capital—Next Three Years	
a. Abundant and inexpensive	
b. Subsidized (i.e. low cost) but restricted	
c. Available but costly	
d. Difficult to get	
7) Availability of Labor—Next Five Years	
a. Plentiful skilled and unskilled	
b. Shortage of skilled labor	
c. Tight and costly	
d. Very tight	

Table 6.3 *(continued)*

Commercial Factors (cont'd)   **Score**

8) Stability of Labor—Next Five Years
    a. Very stable
    b. Active unions but reliable labor
    c. Frequent labor unrest
    d. Continual politically motivated strikes
9) Corporate Tax Level—Next Five Years
    a. Low (income tax less than 35%)
    b. Fair (income tax 35% to 50%)
    c. High (income tax at least 50%)
10) Quality of Infrastructure—Next Five Years
    a. Good services readily available
    b. Adequate but specific shortcomings
    c. Inadequate infrastucture

Monetary-Financial Factors

1) Annual Inflation—Last Three Years
    a. Low (less than 4%)
    b. Moderate (4-8%)
    c. High (8-15%)
    d. Rapid (over 15%)
2) Annual Inflation—Next Three Years
    a. Low (less than 4%)
    b. Moderate (4-8%)
    c. High (8-15%)
    d. Rapid (over 15%)
3) Number of Devaluations—Past 10 Years
    a. No devaluation
    b. One to two devaluations
    c. Weak with predictable manipulations
    d. Three devaluations   2
    e. Continual devaluations   1
4) % of Devaluation—Past 10 Years
    a. Zero   5
    b. Up to 5%   4
    c. 5-10%   3
    d. 10-20%   2
    e. Over 20%   1
5) Currency Forecast—Next Three Years
    a. Consistently strong currency   16
    b. Occasional weakening   12
    c. Weak with predictable manipulations   8
    d. Continual depreciation expected   4
6) Overall Balance of Payments—Next Three Years
    a. Strong with consistent surpluses   12
    b. One to two deficits expected   6
    c. Continuous deficit position   2

Table 6.3 *(continued)*

Monetary-Financial Factors (cont'd)	Score

7) External Debt Position—Next Three Years
   a. Repayment manageable   8
   b. Some repayment difficulties   5
   c. Severe measures needed to avoid default   2

8) Reserves/Imports Ratio—Past 12 Months
   a. Highly favorable (three months or more)   6
   b. Satisfactory (two to three months)   4
   c. Unsatisfactory (less than two months)   2

9) Reserves/Imports Ratio—Next 12 Months
   a. Highly favorable (three months or over)   8
   b. Satisfactory (two to three months)   5
   c. Unsatisfactory (less than two months)   2

10) Convertibility in Foreign Currencies—Next Three Years
   a. Freely convertible   16
   b. Minor restrictions   12
   c. Strict controls on specific transfers   8
   d. Prior approval for all transfers   4

Energy Vulnerability

1) Dependence on Imported Oil for Energy Needs—Next Five Years
   a. Self-sufficient or net exporter   14
   b. Low (1-25% dependent on oil imports)   10
   c. Moderate (26-50%)   6
   d. High (over 50%)   2

2) Import Oil Bill as a Percent of Total Exports—Last Three Years
   a. Minimal (0-5%)
   b. Moderate (6-20%)
   c. Large (21-35%)
   d. Over 35%

3) Development of Alternate Energy Sources—Last Three Years
   a. Oil self-sufficient—no need to develop alternatives
   b. Meaningful progress
   c. Little or no development of alternate energy programs

4) Growth in Use of Alternate Energy Sources as a Percent
of Total Consumption—Next 10 Years
   a. Oil self-sufficient—no need to develop alternatives
   b. High (over 10% of energy consumption)
   c. Moderate (5-10%)
   d. Small (0-5%)

SOURCE: Reprinted from the Business International research report: "STRATEGIC PLANNING FOR INTERNATIONAL CORPORATIONS: Organization, Systems, Issues and Trends" (December 1979), pages 89–93, with the permission of the publisher, Business International Corporation (New York).

Figure 6.5 Haner's Framework for Political Risk Evaluation

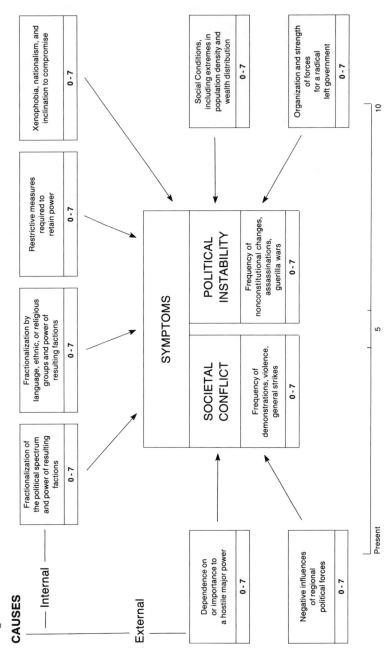

SOURCE: F.T. Haner, "Rating Investment Risks Abroad," *Business Horizons* 22 (April 1979): 18–23. Copyright 1979, by the Foundation for the School of Business of Indiana University. Reprinted by permission.

Figure 6.6 Channon and Jallard's Political Acceptability Profile

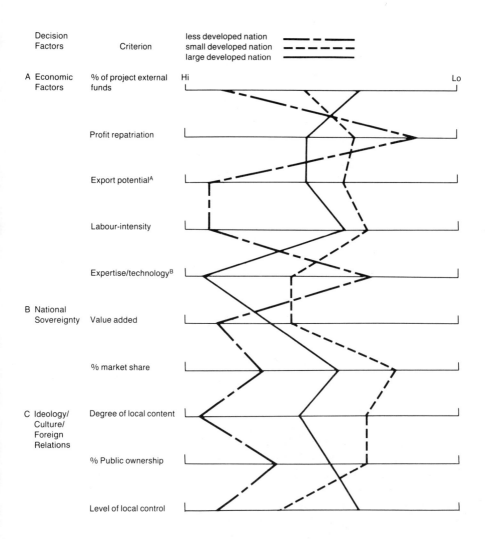

(A) Operationalized in terms of the percentage of profits that
will be reinvested in the host country.

(B) Operationalized in terms of whether the investment introduces
new techniques and additional training of personnel.

SOURCE: Reprinted by permission from Derek F. Channon and Michael Jallard, *Multinational Strategic Planning* (New York: American Management Associations, 1978). Copyright 1978, by Amacom.

ticulars of a proposed project. The criteria can be expanded to include other issues that may be important to the host government. In this manner, the profile may be more sensitive to the interests and policies of the state. The range can be modified to reflect specific values, and different weights can be assigned and scaled into the profile.

In effect, however, the political acceptability profile suffers from the same measurement and scoring problems that are endemic to rating scales. Knowing the actual goals of the government, moreover, is a far more complex problem than it first appears to be. Publicly stated goals may be mere window dressing, while the government pursues other concerns. Goals often are not translated into policies. Governments, particularly those in the Third World, may be unable to implement many policies because they lack the capabilities or are constrained by other factors. Pragmatism, as many avowedly socialist leaders in the developing world have exhibited time and again, often comes before ideology.

The profile is rooted totally in current conditions. There is no attempt at forecasting future developments. The factors and criteria are not directly related to particular political risk events. The political acceptability profile is like a silhouette: it reveals the outlines of political conditions, but it lacks detail and contour.

"Based on the assumption that political risk, while a product of a very complex set of causal factors, can be broken down into measurable components," Dan Haendel, Gerald T. West, and Robert G. Meadow recommend a components approach to political risk.[28] The specific components are more amenable to analysis than the larger problem of political risk. They can be measured and "reaggregated in a form suitable to meet the particular needs of the political risk analyst."[29] This enables the analysts to use what they consider hard, quantitative data, rather than soft measures, such as those employed by the Delphi method.[30]

Haendel, West, and Meadow have constructed a "political system stability index" (PSSI) that ranks the relative stability of Third World countries. The index is comprised of three equally weighted indices: socioeconomic, governmental processes, and societal conflict. The socioeconomic and governmental process indices are based on three and four equally weighted indicators, respectively. The societal conflict index is based on three unequally weighted subindices: public unrest (0.2), internal violence (0.4), and coercion potential (0.4). The subindices are measured in terms of nine indicators. Figure 6.7 illustrates the PSSI.

The PSSI ratings are based on empirical data. The scores are standardized for comparability. The higher the score, the more stable the country's political system is relative to the system of other states. The subindices are complemented by a confidence rating of 1 through 5.

Haendel, West, and Meadow show the feasibility of a components approach in constructing an index of political risk. Since it is drawn solely from empirical data, however, the PSSI focuses on current conditions and does not forecast future

Figure 6.7 Haendel, West, and Meadow's Model for the Formation of the Political System Stability Index (PSSI)

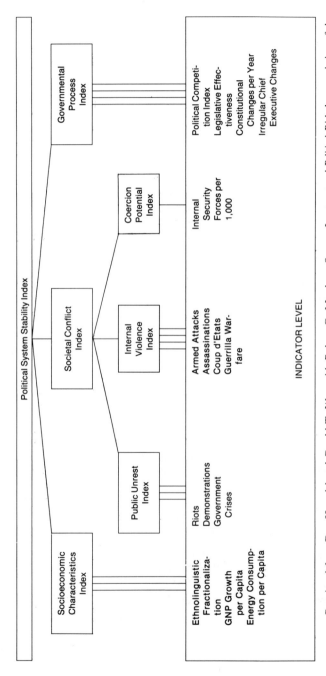

SOURCE: Reprinted from Dan Haendel and Gerald T. West with Robert G. Meadow, *Overseas Investment and Political Risk*, by permission of the publisher. Copyright 1975 by the Foreign Policy Research Institute, Philadelphia.

developments. The authors suggest that the PSSI can be adapted as a forecasting tool. Other than predicting future indicators, however—a procedure that violates their claim to use only hard data[31]—it is left unclear how forecasting is to be accomplished. Although the model is explicit and there is an explanation of the selection of indicators, the dependent variable is poorly conceptualized. Despite a paragraph that tries to justify the use of political stability as the dependent variable, the case is not convincing. The methodology also is too general, with little application to specific situations or to decision making.

The authors explicitly state that the procedure is not applicable to the "United States or developed states in general."[32] This reduces the utility of the PSSI as a comparative tool. However, neither in the elaboration of the relationship between a society and the political system nor in the reasoning on the selection of indicators is the distinction between developed and developing state elaborated.[33] If societies and political systems behave as the authors suggest, then the model should have a universal application, and the exclusion of the developed countries seems unjustified; if developed and developing systems behave differently, the model needs to be advanced in the particular context of the sociopolitical process of the Third World.

The decision tree approach to evaluating political risk produces a series of cumulative probabilities of specified occurrences. Figure 6.8 illustrates an array of hypothetical probabilities associated with the likelihood of a change of government and expropriation of corporate assets within one year. As probabilities are multiplicative and have a summation of one, the decision tree enables analysts to compute the likelihood of a combination of events and situations.

The decision tree is easily adaptable to analysis in a specific context and can be structured to address the particular interests of a firm that is planning to invest abroad. The dependent variables can be conceptualized explicitly to represent the concrete concerns of the company involved. A number of dependent variables can be combined in an expanded tree that computes the probability of multiple political risk events. Alternative corporate decision choices can be directly related to likely scenarios. Moreover, the decision tree yields results that are easily integrated into the corporate decision-making process. Projected cash flows, for example, can be multiplied by the probabilities of alternative outcomes to determine their present value.

The decision tree also has serious shortcomings. Having no model of political risk, it does not tell us which variables, events, and policies should be included in the analysis. Intuitive selection of independent variables may cause us to overlook some issues and overconcentrate on others. Although the decision tree gives us a statistical procedure for combining relevant probabilities, it does not tell us *how* to calculate those probabilities. The assignment of subjective probabilities means that the results will be subjective; no matter how scientific the procedures for combining probabilities (data, inputs) are, if the probabilities are subjective, the outcomes also will be subjective. This problem might be overcome if the decision tree were combined

Figure 6.8 Hypothetical Decision Tree—Probability Analysis, Year One: Government Change and Expropriation of Project (.000 = probability)

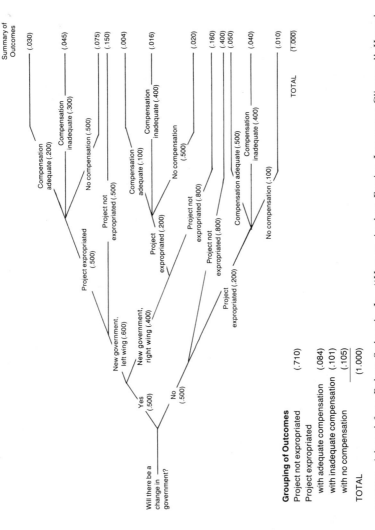

SOURCE: Adapted from Robert Stobaugh, Jr., "How to Analyze Foreign Investment Climates," *Harvard Business Review* 47 (September–October 1969): 100–108.

with a model of political risk and an analytical procedure for evaluating probabilities. Devising a means to forecast such probabilities, however, is a major task in and of itself and has yet to be accomplished.

Frost and Sullivan's World Political Risk Forecast (WPRF) combines an input-gathering procedure with a methodology to project probabilities. Arguing that "those approaches that make heavy use of quantitative indicators and statistical models based on aggregate data are the least promising," William D. Coplin and Michael K. O'Leary, the designers of the WPRF, suggest that "more useful approaches are based on the use of surveys of specialists that elicit both narrative and numerical responses."[34] The specialists are used as "sources of information about current conditions," not as forecasters;[35] the information is processed and a forecast is produced by means of the Prince system.

Political risk events or issues to be forecast must be selected. The specialist then identifies those political actors—decision makers, individuals, groups, institutions, or organizations—that will play a role in determining whether the event occurs. The expert scores the actors in terms of their "orientation" (for, against, or neutral); "certainty" (of their orientation); "power" (how influential they are in making the relevant decision); and "salience" (how much importance they attribute to the event). Orientation is scored to indicate direction (+, 0, −); the other three variables are scored on a range from 1 (little/none) through 5 (extremely high). The scores of each actor are multiplied. The sum of the positive scores (including half of each neutral score) is divided by the absolute value of the sum of all scores. The result is the probability of the event occurring.

The standardized WPRF questionnaire includes five issues: regime change, political turmoil, restrictions on international business, trade policies, and economic policies. Prince charts are used in each section except economic policies. Figure 6.9 is a sample Prince chart for forecasting restrictions on international business. The forecasts are made for an 18-month period and are supplemented by estimates of likely changes, as perceived by the expert who completes the questionnaire. The expert also makes a more general five-year forecast. The quantitative procedure is supported by a descriptive/qualitative report that addresses the overall political situation and the actors incorporated in the Prince charts.

The standardized format of the WPRF ensures a measure of comparability across countries and permits the monitoring of national changes over time. The WPRF addresses a number of dependent variables, and the model can be used to forecast other events as well. Probability results, as we mentioned earlier, are relatively easy to incorporate in decision making.

Although the events or dependent variables can be adjusted somewhat, industry-, firm-, and project-specific concerns are ignored in the WPRF. The national probability of political turmoil or business restrictions may be of only vague relevance to any particular firm.[36] This reduces the utility of WPRF probabilities to a

Figure 6.9 Frost and Sullivan WPRF Sample Prince Chart

I. PRINCE CHART OF SUPPORT FOR OR OPPOSITION TO RESTRICTIONS ON INTERNATIONAL BUSINESS.
(An extremely certain supporter of restrictions would be an ideologically based actor which opposes all forms of international business; an extremely certain opponent of restrictions would welcome all foreign investment under extremely liberal conditions.)

CURRENT ESTIMATES

Orientation	CERTAINTY, POWER, SALIENCE
+ Supports	1 LITTLE/NONE
	2 SLIGHT
N Neutral	3 MODERATE
	4 HIGH
– Opposes	5 EXTREMELY HIGH

REVIEWER
Either indicate with a check you agree with estimate or forecast or make changes with a different color. You may also add actors.

ACTOR

ORIENTATION × CERTAINTY × POWER × SALIENCE = PRINCE SCORE

x	x	x	=	☐
x	x	x	=	☐
x	x	x	=	☐
x	x	x	=	☐
x	x	x	=	☐
x	x	x	=	☐
x	x	x	=	☐
x	x	x	=	☐
x	x	x	=	☐
x	x	x	=	☐
x	x	x	=	☐
x	x	x	=	☐

Check the box above the number which best represents your estimate of the CURRENT LEVELS OF RESTRICTIONS ON FOREIGN BUSINESS.

0	1	2	3	4	5	6	7	8	9	10
	Low Restrictions				Moderate Restrictions			High Restrictions		

Check the box above the number which best represents your estimate of the LIKELIHOOD OF NEW RESTRICTIONS WITHIN THE NEXT 18 MONTHS.

0	1	2	3	4	5	6	7	8	9	10
Very Unlikely		Unlikely			Uncertain		Likely		Very Likely	

SOURCE: Reprinted from Frost and Sullivan Inc. Copyright 1982. Reprinted by permission.

143

firm's decision-making process. As in the decision tree, *how* the expert arrives at the rating for the variables is undefined. This is subjectively determined, and, as Coplin and O'Leary admit, "the Prince system does not produce any better forecasts than the quality of the input from knowledgeable observers."[37]

Unlike most other analytical methods for evaluating political risk, the WPRF uses an explicit model of the political process. Political events are seen as the product of the interaction of key political actors. The Prince system borrows some of the logic of decision-making analysis, an approach that is popular in the study of foreign policy. It recognizes institutions and oganizations as decision makers and assumes a smooth, pluralistic political process.[38] It is questionable, however, whether the political processes function as rationally as the Prince system implies.[39] Focusing on political actors is important insofar as they make the decisions. Political decisions are not made in a vacuum however; they occur in a complex social, political, and economic environment. The WPRF takes this environment into account only indirectly, in terms of how it affects the choices of political actors and in the background reports. The assumption that outcomes are simply the result of the orientation, certainty, power, and salience of the relevant actors, moreover, both is an erroneous conceptualization of the policy-making/implementation process and glosses over the subjective measurement problem.

Perhaps the most specific methodology for analyzing political risks in the oil industry is the ASPRO/SPAIR (Assessment of Probabilities/Subjective Probabilities Assigned to Investment Risks) procedure designed by Shell Oil. ASPRO/SPAIR, which has been the model for a number of other analytical techniques (including that of Bunn and Mustafaoglu) is designed to address the particular political concerns of firms that are investing in petroleum projects abroad. The premises of the project under consideration are stated explicitly:

1. *the oil firm has or is negotiating an agreement for exploration and development of petroleum;*
2. *the contract is seen as fair by both the firm and host;*
3. *the project will prove successful; and*
4. *the amount of crude discovered will be sufficient to have "a significant economic impact" on the host for the next decade.*[40]

Given these premises, "political risk is defined as the probability of not maintaining the described contract during a 10-year time span in the face of changing economic and political circumstances."[41]

Political risk is divided into nine components, which are grouped in two categories. The larger category (seven components) involves changes that result in an inadequate return on an investment *in the host country*. The smaller grouping includes the risks of restrictions on the movement of funds and oil *out of the country*.[42] Figure 6.10 presents the risk components of the ASPRO/SPAIR model. The economic,

Figure 6.10 ASPRO/SPAIR Components of Political Risk

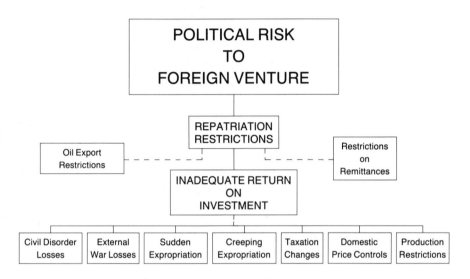

SOURCE: Reprinted by permission from C.A. Gebelein, C.F. Pearson, and M. Silbergh, "Assessing Political Risk of Oil Investment Ventures," *Journal of Petroleum Technology*, May 1978, pp. 725–730. Copyright 1978, SPE-AIME.

political, and social factors that are expected to determine the occurrence of the component events are dichotomized and included under each component, which is reformulated in the form of a proposition. The propositions (for example: "Tax rates on oil revenues are increased by the host country government so that, if viewed in retrospect by the oil company, after-tax revenues would not have justified the original investment"[43]) are evaluated in terms of the relevant factors. Each factor is reviewed independently. Although the factors are treated as a complete set, an expert can add variables he believes should be included.

A panel of experts, composed of individuals with different academic and national backgrounds, is used. The experts are asked to evaluate each component and factor in light of alternative (pessimistic/optimistic) scenarios. This generates a type of range of estimates. The expectation is that the real probability lies between the best estimates (the panel mean) for the optimistic and pessimistic scenarios.[44] To ensure feedback and allow the experts to decide whether their positions are accurately reflected in the quantification procedure, the results are shown to the experts.

The probabilities are expressed on a qualitative scale, ranging from strongly refuting and weakly refuting, to neutral, to weakly supporting and strongly support-ing. The level of support is translated into quantitative terms by assigning values of 10 percent, 30 percent, 50 percent, 70 percent, and 90 percent, respectively. The probabilities of a particular event, based on the separate factors, are combined for each expert. The results for the various experts then are aggregated, taking con-fidence levels into accounts. The results are plotted, showing the probability density and best estimate (mean) of the distribution of the panel's aggregate forecast for both the optimistic and pessimistic scenarios. Figure 6.11 is a sample graph for country X.

ASPRO/SPAIR is structured to be firm-and project-specific. The dependent variables are conceptualized clearly and expressed in operational terms that are directly related to the interests of oil companies operating abroad. The propositions can be modified easily to reflect the particular concerns of corporate executives. As in the WPRF framework, the experts do not directly predict the probability of risk events; rather, they evaluate variables, and the probabilities are determined by statistical procedures. The use of alternative scenarios and best estimates provides decision makers with a range of probabilities based on a high level of confidence.

Despite its sophistication, ASPRO/SPAIR is not without its problems. The assignment of values ranging from 10 percent to 90 percent to qualitative terms seems arbitrary and static. Every factor is weighted equally. This may reduce the problems of subjectively weighing different factors, but the assumption that each factor is of equal importance in determining the truth of a proposition (the probabil-ity of a political risk event) is grounded neither in theory nor in empirical investiga-tion. From a practical viewpoint, moreover, the procedure is somewhat arduous and time consuming, although the cost may well be offset by the corporate specificity of the product.

The various approaches to analyzing political risk that we have summarized here vary widely in terms of their rigor and specificity, as well as in their costs of completion. Each technique has some advantages and each has shortcomings. The methodology that is best depends on the particular circumstances involved.

Basing a decision on a rating scale of some kind would seem imprudent. The scales are useful, however, in monitoring changes in the investment environment and pointing to those variables that are important. The techniques tend to be inex-pensive and easy to implement, not requiring an elaborate model, a verification pro-cess, or a complete panel of experts.

The other approaches are more sophisticated and specific in application. They are also more complex, however, and often costly to complete—in terms of material and human resources and time. For simple, ongoing monitoring of a situation, these costs may be prohibitive; for an initial investment decision, however, the costs may be more than worth the rewards earned in gaining a risk forecast that is better suited to the needs of decision makers.

Figure 6.11 ASPRO/SPAIR Probability of Taxation Changes
for Country X, 1982–1987

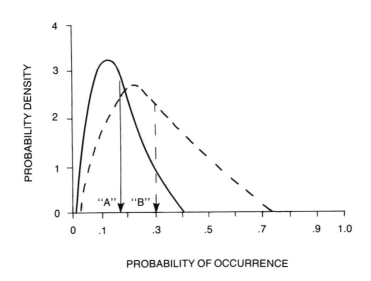

PROBABILITY OF OCCURRENCE

"A" = BEST ESTIMATE, "OPTIMISTIC" SCENARIO
"B" = BEST ESTIMATE, "PESSIMISTIC" SCENARIO

*The aggregate panel judgement about the probability that the host country tax rates
will be increased over the next ten years to the point that, if viewed in retrospective,
after tax revenues would not have justified the original investment.

SOURCE: Reprinted by permission from C.A. Gebelein, C.F. Pearson, and M. Silbergh, "Assessing
Political Risk of Oil Investment Ventures," *Journal of Petroleum Technology*, May 1978, pp. 725–730.
Copyright 1978, SPE-AIME.

Combining approaches and modifying them to cater to specific needs and circumstances may be useful for gaining multiple perspectives. Over time, the company can refine the methodologies and discard those that actually prove the least useful to the firm. Those techniques that perform the best and have the best "track record" can be retained, along with one or two contrasting approaches for comparison. Although this approach is attractive from the point of view of best matching a company with a political risk assessment technique, the associated costs of such an option may be too high. The way firms use political risk analyses leads us to the issue of managing political risk, which is addressed in the next chapter.

Notes

1. Stephen J. Kobrin, "Political Assessment by International Firms: Models or Methodologies," *Journal of Policy Modeling* 3 (1981): 255.
2. Howard Johnson, *Risk in Foreign Business Environments: A Framework for Thought and Management* (Cambridge, Mass.: Arthur D. Little, 1980).
3. Robert T. Green, "Political Structure as a Predictor of Radical Political Change," *Columbia Journal of World Business* 9 (Spring 1974): 29–35.
4. David A. Jodice, "Sources of Change in Third World Regimes for Foreign Direct Investment, 1968–1976," *International Organization* 34 (Spring 1980): 177–206.
5. Ibid., pp. 189–198.
6. Ibid., p. 205.
7. Franklin R. Root, "Analyzing Political Risks in International Business," in *The Multinational Enterprise in Transition*, ed. A. Kapoor and Phillip D. Grub (Princeton: Darwin Press, 1972), p. 359.
8. Ibid.
9. Johnson, *Risk in Foreign Business Environments*, p. 11.
10. See Leonard Binder et al., *Crises and Sequences in Political Development* (Princeton: Princeton University Press, 1971).
11. Johnson, *Risk in Foreign Business Environments*, pp. 10–11.
12. Ibid., p. 10.
13. D.W. Bunn and M.M. Mustafaoglu, "Forecasting Political Risk," *Management Science* 24 (November 1978): 1557–1567.
14. Ibid., pp. 1558–1559.
15. Ibid., p. 1562.
16. See Ted Gurr, *Politimetrics: An Introduction to Quantitative Macropolitics* (Englewood Cliffs, N.J.: Prentice-Hall, 1972).
17. P. Nagy, "Quantifying Country Risk: A System Developed by Economists at the Bank of Montreal," *Columbia Journal of World Business* 13 (Fall 1978): 141.
18. Kobrin, "Political Assessment," pp. 255–256.
19. R.J. Rummel and David A. Heenan, "How Multinationals Analyze Political Risk," *Harvard Business Review* 56 (January–February 1978): 69.
20. Business International research report. *Strategic Planning for International Corporations* (New York: Business International Corporation, 1979), p. 85.
21. Ibid., p. 86.

22. Dan Haendel, *Foreign Investments and the Management of Political Risk* (Boulder: Westview, 1979), p. 99.
23. Kobrin, "Political Assessment," p. 261.
24. F.T. Haner, "Rating Investment Risks Abroad," *Business Horizons* 22 (April 1979): 19-21.
25. Ibid., p. 21.
26. Derek F. Channon and Michael Jallard, *Multinational Strategic Planning* (New York: American Management Associations, 1978), pp. 243-245.
27. Ibid., p. 243.
28. Dan Haendel, Gerald T. West, and Robert G. Meadow, *Overseas Investment and Political Risk* (Philadelphia: Foreign Policy Research Institute, 1975), p. 58.
29. Ibid.
30. Ibid., p.62.
31. Ibid., p. 61.
32. Ibid., p. 67.
33. Ibid., pp. 61-62, 79-83.
34. William D. Coplin and Michael K. O'Leary, "Political Analysis in the Forecast of Oil Price and Production Decisions," paper presented at the Electric Power Research Institute Fuel Supply Seminar, Memphis, 8-10 December 1981, p. 12.
35. Ibid., p. 16.
36. Kobrin, "Political Assessment," p. 263.
37. Coplin and O'Leary, "Political Analysis," p. 23.
38. Kobrin, "Political Assessment,"p. 262.
39. See, for example, Graham T. Allison, *Essence of Decision: Explaining the Cuban Missile Crisis* (Boston: Little, Brown, 1971), or John D. Steinbruner, *The Cybernetic Theory of Decision* (Princeton: Princeton University Press, 1974).
40. C.A. Gebelein, C.F. Pearson, and M. Silbergh, "Assessing Political Risk of Oil Investment Ventures," *Journal of Petroleum Technology*, May 1978, p. 726.
41. Ibid.
42. Ibid.
43. Ibid., app. B, pp. 729-730.
44. Ibid., p. 728.

7

Corporate Management of Political Risk

Corporate Decision Making and Political Risk Revisited

The decision to invest in petroleum and natural gas production abroad is simultaneously more straightforward than the textbook theory of direct foreign investment and more complicated than the apparent simplicity of weighing costs and benefits. The primary concern of the firm is making a profit. (Although often described as such, it is doubtful that corporations can be described as profit *maximizers*.) Closely related to this main objective are the related interests in assuring adequate crude and gas supplies and in controlling those supplies. At times, the pursuit of access to and control over sources of supply may eclipse the profit motive in the short term, but the companies' main reason for existence is making money.*

   Foreign investment opportunities are enticing (or not) because of the promise of profits and/or sources of supply. Higher revenues may be realizable from lower costs for crude or gas (whether because of lower production costs or a less onerous fiscal

---

*Nocs have the dual concerns of furthering state policy goals and earning profits. Consumer state nocs may be more interested in assurance of supply as a national concern than their private company counterparts are. Despite their relationship to their respective governments, consumer and producer nocs must pay careful attention to profitability. In this sense, nocs share the same concerns typical to all oil companies; however, nocs are more likely to be subject to policy guidelines that may or may not take profitability into account.

regime), the opportunity to enter new markets, transportation cost differentials, or increased volumes of sales. Diversifying the sources of supply helps insulate firms from overreliance on any one source, thus protecting it from an interruption in imports from a supplier. A transnational or multinational production network can be helpful in balancing global and regional supply and demand in the most cost-effective manner and in managing a vertically integrated firm. Quality and price differentials of different producers can be matched to needs and exploited to net further rewards.

The assertiveness and market power of oil-exporting countries have eroded much of the control and flexibility traditionally exercised by the oil TNCs. The companies are less able to coordinate liftings from the various sources than they were prior to the "OPEC revolution." Direct investments have been modified via expropriation or participation, often to be replaced by minimal equity investments or nonequity agreements. The near-adversarial relationship of the international marketplace in crude (and, to a lesser extent, in natural gas), has intensified corporate sensitivity to profits and assuredness of supply.

What is important to the oil and gas companies is not the bundle of legal rights that traditionally accompanied ownership and foreign direct investment. Rather, the concerns are with how a project or agreement—however structured—will yield profits to the firm or provide it with sufficient supplies. Companies in the crude and natural gas industry have proved themselves amenable to a wide variety of types of investment agreements, provided that the relationship promises to be consonant with corporate interests.

The presence of political risks and uncertainty about the future complicates the theoretical simplicity of calculating costs and benefits. The conditions and investment regime that prevail at the outset of a proposed project may be inviting and may ostensibly justify an investment commitment. What is the likelihood, however, that the project will operate under those same terms throughout the life of the agreement? What are the alternative scenarios and probable returns? What about unforeseen developments? The guesses, judgments, assumptions, or forecasts of decision makers on these and numerous other questions are of vital importance in evaluating the worth of an investment abroad (and at home). Unless risk is ignored, various unknowns must be incorporated into the cost-benefit calculations.

The forecasting of political risk is an essential link in the decision-making chain. Given an investment opportunity, an educated decision on whether or not to pursue a project rests on a proper understanding of the variables that will render an investment profitable or not. Political risk is one of the more elusive variables and perhaps the most difficult to measure. It needs to be treated explicitly as a factor in decision making, rather than assumed or given.

Figure 7.1 presents Dan Haendel's conceptualization of the decision-making process for foreign investments. Evaluating risks actually is only one step in the larger corporate goal of managing political risks. The management of political risk is the

Figure 7.1 Foreign Investment Decision Flow

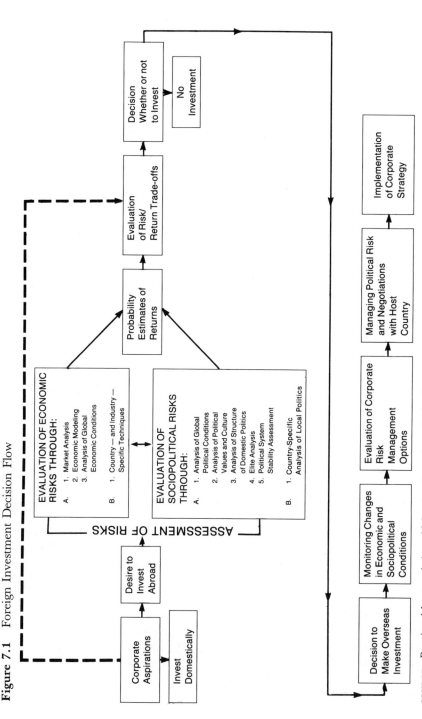

SOURCE: Reprinted by permission of Westview Press from *Foreign Investments and the Management of Political Risk* by Dan Haendel. Copyright 1979 by Westview Press, Boulder, Colorado.

umbrella term encompassing corporate concerns with such risk and integrating it into decisions and policies. Management is an ongoing process. Internally, managing political risk entails monitoring political changes, keeping abreast of the potential consequences changes may have for the company, and, if necessary, adapting company policy. Externally, the management function is representing company interests and concerns about changes to the host government as well as maintaining the host-firm relationship. At times, the company also may need to manage the political risks associated with activities by the home government.

Some firms stop short of completing the decision flow as modeled by Haendel. Instead of following through with the monitoring and management processes, a company may see the political risk function as ending with the decision on whether or not to make an investment. This is characterized by Stobaugh as the go–no go approach.[1] Using a checklist or some other rudimentary approach, the firm classifies the prospective host as falling either within or outside the band of acceptable political risk. Projects in countries that are classified as *too risky* are rejected automatically. Projects in countries that are judged acceptable will be pursued or rejected depending on the current financial, economic, and geological prospects.

The go–no go approach is simple but superficial. It fails to take into account exactly what political risk analyses strive to reveal: the chances of changes—both positive and negative—at some time in the future. The not surprising tendency of this approach to decision making is to avoid investments in developing countries in favor of projects in developed countries and to reinvest in those countries where the firm has existing operations. Assumptions about the relative safety in more-developed countries and in nations where there are ongoing projects and the go–no go approach tend to be mutually reinforcing. The opposite inertia also prevails; once judged unsafe, nations may be disqualified automatically, regardless of changes.

Use of the go–no go approach is a step above ignoring political risks. Few firms would be so brazen and unwise as to disregard political developments. Many do treat political issues implicitly, however, and do not maintain any sort of formal political risk evaluation procedure. This is less common in the extractive sector than in other areas of the economy. The more typical and relevant problem, which will be addressed later in this chapter, is the failure to integrate political risk in the decision-making process. Brilliant analyses are merely academic if they are not considered properly in decision making.

A more common approach than the go–no go types of decisions is to assign a higher discount rate or to require a higher rate of return on investments in countries that are not considered good risks. Adding a premium for political risks may be more logical than a go–no go decision, but the addition of premiums does not directly address the problems of political risk. In a sense, this is a go–no go decision modified to account for risks by requiring larger returns. The assignment of premiums usually is arbitrary. The higher rate of return demanded on projects that are seen as more risky might mean that the company effectively disqualifies itself from promising invest-ment opportunities. Political risk, moreover, is dynamic and cannot be reflected fully in a static premium. Fluctuating political fortunes cannot be smoothed out by unilaterally assigning a risk premium, such as an additional 10 percent, to traditional target profit levels. As in the go–no go approach, risk premiums usually carry a heavy bias in favor of developed countries. The target rate of return—15 percent, for example—often is based on and applied to projects in OECD countries, while developing states are slated for a higher-risk pool.

Host governments are reluctant to recognize the legitimacy of being required to pay a political risk premium to attact investment. No regime will admit to the likelihood that political changes will negatively affect foreign firms, and they are even less willing to address such issues as political instability. Premiums often are translated into what the host government sees as excessive profits. This might pro-voke political action against a foreign company, as charges of economic exploitation may arise. Premiums reinforce the sense of discrimination against the Third World by the developed states—a sensitive nerve in the differences embodied in the North-South confrontation—and may backfire against the transnational firm.

Despite the problems, premiums are commonly employed by many firms. Because they are uncertain about the future, companies often prefer to try to protect themselves with higher returns. As in the go–no go approach, levying premiums removes political risk as an explicit factor in decision making and ignores the concept of political risk management. Attempting to control for political risks by adjusting discount rates or adding premiums may skew investments and even may induce some of the risks that the firm wants to avoid.

To be useful, political forecasts have to be used. Although this might seem to be stating the obvious, it points to a common problem: political risk analyses often are not part of decision making and therefore have little if any impact on decisions. There are a number of reasons why forecasts are often ignored, including the biases of corporate management, the nonapplied approach of many analysts, and structural obstacles stemming from corporate organization. The bottom line, one petroleum in-dustry analyst has concluded, is that despite the variety of methods in evaluating political risk,

> one point rises above all of them: that regardless of the work being done, the factors
> studied, the signals being weighed, all of this work is largely ignored, or not taken into
> consideration, and that decisions are finally based on the hunches, or particular pre-
> judices of the individual.[2]

Political upheaval is not new to the petroleum industry. More than other areas of the world economy, oil companies have experienced the losses and fortunes of political changes. This has been an experience with a history, leaving the companies with both political scars and managers that are well aware of how a turn of the political tide might affect company interests. There is a tendency, however, to substitute political sensitivity on the part of decision makers for more formal analyses. Considering themselves politically astute and having practical experience, some oil industry executives are not receptive to lectures on political risk or to analyses by individuals who lack their years of exposure to the industry and countries involved. Although experience is an important component in the decision-making process, however, it is no substitute for analysis. It would be preferable for the experience of corporate management to be integrated into the analysis.

Conversely, there are decision makers who are convinced that political events are exogenous factors that cannot be accounted for by the company. Despite the industry's political exposure, many executives are not politically atuned and downplay political concerns. To some people, political risk analysis is seen as getting involved in politics, which is taboo. Others have no faith in the forecasts and prefer following their own judgments or not considering risk rather than relying on evaluations in which they place no faith.

In many instances, the political risk analysts must shoulder a large share of the blame for having their work go unnoticed by corporate decision makers. For corporate purposes, academic exercises are useless. Analyses should be designed for practical applications. Management may not be convinced of the utility of risk forecasts because the analyses are not oriented to the company's particular interests or lack applied value. Firms must make decisions according to specific conditions; if the analysis is not directed to those conditions, it may not be useful. Framing the issues in terms of the specific interests and concerns of the company is essential. The conclusions or recommendations of forecasters, moreover, should be expressed in a manner that makes good business sense. Academic exigencies may be important to theory building or support work, but they are of little interest to corporate management. Analysts must be careful not to lose sight of the practical needs of the company, a point that sometimes has not been appreciated.

Perhaps the most important major obstacles are structural in nature. Unlike other corporate positions, there is no domestic analog or parallel to the political risk analyst. The closest roles might be handling government affairs and lobbying, neither of which prepares people to evaluate political conditions abroad. As a field that has come to be recognized in its own right only in the last few years, political risk analysis lacks a traditional role in the corporate structure. This deprives the analyst of a natural mooring in the firm and a pad from which to project and draw support. This problem is aggravated by the interdisciplinary nature of risk evaluation. Although it is often subsumed in the strategic planning, financial, or exploration arms of the company, political risk analysis has ben housed in virtually every conceivable division in various firms. The responsibility for political evaluation is often fragmented among individuals in different areas of the company, each looking at a part of the picture and acting in mutual ignorance of one another.

Rather than having an identifiable political risk group that employs full-time professional staff, is headed by a senior person, and reports directly to top management, many firms are organized in a manner that weakens risk analysis and makes it less probable that such analyses will be heard by decision makers. Very often political risk is the part-time responsibility of individuals that were trained to do—and do—other work. These people may be spread over the numerous divisions of a firm, precluding even the most minimal effort at coordination and teamwork. Often, no one is in charge of political risk assessment. Because it does not have a separate division and is handled by an assortment of people on a part-time basis, moreover, senior level personnel are rarely involved in political risk evaluation. Hence, there is no respected high-level spokesperson who has the ear of decision makers to make certain that they are properly informed of any risk forecasts. There is no corporate constituency that focuses on or even has an ongoing professional interest in risk evaluation. It seems that, at almost every turn, the typical corporate structure militates against the use of political risk analyses in the decision-making process.

Although they are substantial, the structural obstacles to better integrating the risk analysis function into corporate decision making are not insurmountable. A separate political risk unit composed of full-time professionals should be organized within the firm. This ensures an identifiable cadre of people who will pursue political risk analysis, as well as establishing a focus of responsibility for such work. This unit or group would centralize the risk analysis function from the gathering and organizing of data and information through the completion of political risk analyses. This would facilitate coordination. The group should monitor political developments on a regular basis and should make recommendations to top management about particular decisions and risk management.

The unit head should be at the managerial level and should have direct contact with corporate executives. The staff should include people who are familiar with the methodologies of political risk analysis, are trained in international politics and political economy, have international political and corporate experience, are area specialists (when feasible), and are intimately acquainted with the company and its operations. The head of the group should fall in the last category. This insures a respected spokesperson for the unit, who will be heard by top management and will provide a control to be certain that all the forecasts and related documents are properly directed toward the position and interest of the company. The diversified backgrounds of the staff would be appropriate to the interdisciplinary nature of inquiries into political risk. The use of area specialists, general political risk analysts, and people who are knowledgeable about company operations and financing would yield a well-balanced mixture of expertise. This would also provide the needed "translation function"[3] between experts in politics and those who know the company inside and out. As a separate unit, the group would be autonomous, which would protect its institutional integrity and ensure that it is not co-opted by any particular corporate interests.

The sources of information used could be expected to change significantly under the restructured group. Traditionally, internal sources of information have been by far the most important to corporate political risk analysts, followed by banks. A survey by the Conference Board reveals that 75 percent of the individuals responsible for corporate (extractive and nonextractive) risk evaluation cited the managers of subsidiaries as the first or second (out of five) most important source of information.[4] Regional managers were cited by 69 percent, followed by headquarters personnel by 65 percent. Of the other sources mentioned, banks (45 percent) were the only one to exceed half of the numbers using company sources.

Although sharing of in-house information is important—even essential—to the risk analysis group we designed, outside sources need more attention. Consultants, cited by slightly more than one-quarter in the Conference Board study, would probably be used more extensively and efficiently by our group to fill gaps in information, approach, and expertise, as well as to provide a second opinion. Academics, which were mentioned by less than 10 percent of those interviewed, also would be used more frequently. The unit we designed would have a better-established network of information and outside sources to draw upon than would the part-time staff that is common to most firms.

Establishing a permanent unit would not instantly solve all the problems associated with political risk analysis and the integration of risk forecasts in corporate decision making. It would provide solid footing, however, from which the group could address the conceptual problems in analyzing risks and the structural and other obstacles to the use of forecasts by top management.

Accounting for Political Risk in Cash Flow Projections

Commercial profitability is a prerequisite for any investment, with the partial exception of government expenditures for public goods. Essentially, an investment is evaluated in terms of an educated projection of costs and returns. Capital investment and the costs of inputs are compared with the anticipated cash flow—the amount and timing of revenues from future sales—of a project to assess its profitability.

Measuring the commercial value of a project can be a difficult task, however. A number of geological and economic variables must be estimated into the future. The lengthy planning horizon means that an oil and gas project will be prey to changes in market conditions over a number of years. The majority of costs are committed in advance of production—perhaps years before the initial investment can be recovered. Changing supply and demand forces must be anticipated and incorporated into the projected cash flow profile. Errors in assumptions may distort the cash flow projection and give a false impression of the value of an investment.

Analyzing the value of an investment requires that future cash flows be discounted to reflect the time value of money. Money has a time value by virtue of its ability to be reinvested and earn future profits. Because of this, a given sum, such as $100, is worth more today than it will be worth next year or ten years from now. Continuously compounding interest at 15 percent annually, a rate that is common to the financial markets at this time, means that today's $100 is worth $116 next year and $448 in ten years. Conversely, discounting next year's $100 at the same rate—that is, measuring the declining future value of money compared to the interest rate applicable to current funds—yields only $86 in today's terms. A cash inflow of $100 next year, therefore, is worth $86 today. By discounting future cash flows, an investment can be analyzed in terms of the current value of future earnings.

Having reduced the cash flow profile to a common denominator—current dollars—a company can gauge the profitability of a project. Although there are numerous ways to measure profitability, the most commonly employed practices evaluate investments in terms of the payback period, the internal rate of return, or the net present value.

The payback period is simply the length of time before a project will recover the initial investment. A faster payback period means that the company recaptures its initial outlay in a shorter span of time, whereas a longer payback period means a more protracted period of time until the initial investment is recouped. The payback period is computed by dividing the investment (outflow) by the projected annual cash flow. Assuming a regular annual cash flow of $25 million, for example, the payback period on a field that costs $150 million to develop would be six years. It is only after payback that the firm earns profits. An accelerated payback period, which can be provided by quicker depreciation and amortization write-offs, a tax holiday, or other accounting fixes, means that a firm can begin to earn profits more quickly than otherwise would be the case. Once payback has been attained, the company has crossed an important threshold; although risks remain important, the threat is with respect to how large profits will be, rather than whether or not there will be any profits.

The internal rate of return refers to the rate that renders the present value of a project zero. This is the point at which the value of future cash flows is equal to the initial investment. The rate of return computes the profitability of an investment as a percentage return on the initial investment. A rate of return of 20 percent, for example, indicates a higher rate of profitability than a return of 15 percent. The acceptance of a project would be determined by comparing the project's estimated rate of return with the firm's minimum acceptable rate of return. The internal rate of return method has fallen largely into disuse and is rarely used in the extractive sectors.

The discounted net present value—perhaps the most commonly employed tool for gauging the worth of an investment—compares the costs of an initial investment with the sum of the discounted cash flow. In our earlier example, if the present value of the sum of future cash flows were $200 million and the investment were $150 million, the net present value would be $50 million. Cash flows are discounted at the firm's cost of capital, which, under perfect competition, equals the market rate of interest. Because of imperfect financial markets and competition, as well as budgetary constraints, the company's cost of capital is usually slightly higher than the market interest rate. For the sake of simplicity, let us assume that the firm's cost of capital is equal to the prevailing interest rate. A positive net present value means that the investment will yield returns in excess of the costs incurred by raising the necessary capital (and greater than the revenues that would be generated by placing an amount of money equal to the investment in an account earning the market rate of interest). In the absence of alternative investment opportunities, a project with a positive net present value should be approved. Conversely, a negative net present value indicates that capital costs exceed real returns and that the firm's money would be better spent elsewhere or should be left in an interest-bearing account rather than invested in the project under consideration.

Traditionally, cash flow evaluations ignored political factors. Accountants have tended to treat the initial investment agreement as a given, to plug in the relevant projections of costs and revenues, and to compute cash flows. Political variables, however, have proved to be important determinants of annual cash flows and of the profitability of hydrocarbon investments. Prices and output, two key economic variables in determining sales revenues, are as much a function of political decisions as they are of market forces. Taxes, royalties, depreciation, and accounting procedures—all of which are central to measuring profitability—are known variables at the time an agreement takes effect. The assumption that these variables will remain unchanged, however, ignores the political dynamics of foreign investment agreements. Projecting future cash flows strictly in terms of the original investment regime, therefore, fails to capture the potential or probable impact that political developments can have on a project's profitability. In effect, moreover, failure to integrate political risks in projected cash flows is tantamount to assuming an ideal political environment in which political developments will not affect the investment regime of a project.

Table 7.1 illustrates how changes in the fiscal regime can alter the returns captured by an oil or gas firm operating abroad. A simplified hypothetical income statement has been constructed to show how the company's net income is computed under eight alternative fiscal arrangements. Case A is indicative of the tax and royalty burden to which a firm might be subjected in a project in a developing country that is not a proved oil producer and relies on imported petroleum.

If commercial fields are found and developed and the country becomes a net exporter, the host government might press for a revised fiscal regime. Increased royalties, as case B reveals, drastically reduce the companies' net income. Because royalties are treated as above-the-line operating expenses, they have a more severe impact than income taxes on corporate profits. Leaving the income tax rate unchanged and raising the royalty from 12.5 percent to 16.67 percent reduces net in-

Table 7.1 The Effects of Royalty and Income Tax Changes on Hypothetical Net Income from a Project ($000)

	A 12.5% royalty, 50% income tax	B 16.67% royalty, 50% income tax	C 12.5% royalty, 55% income tax	D 12.5% royalty, 60% income tax	E 16.67% royalty, 60% income tax	F 16.67% royalty, 65% income tax	G 16.67% royalty, 80% income tax	H 20% royalty, 85% income tax
Gross revenues	950,000	950,000	950,000	950,000	950,000	950,000	950,000	950,000
less Royalties	(118,750)	(158,365)	(118,750)	(118,750)	(158,365)	(158,365)	(158,365)	(190,000)
Net revenues	831,250	791,635	831,250	831,250	791,635	791,635	791,635	760,000
less Operating Costs	(768,000)	(768,000)	(768,000)	(768,000)	(768,000)	(768,000)	(768,000)	(768,000)
Gross income	63,250	23,635	63,250	63,250	23,635	23,635	23,635	(8,000)
less Income taxes	(31,625)	(11,818)	(34,788)	(37,950)	(14,181)	(15,363)	(18,908)	—
Net income	31,625	11,818	28,462	25,300	9,454	8,272	4,727	(8,000)

come in case B to approximately 37 percent of what it was under the initial agreement. Leaving the royalty constant, changes in taxes have an important but far less dramatic effect on earnings: with royalties unchanged, a jump in taxes from 50 percent to 60 percent reduces net earnings to 80 percent of their previous level.

Let's assume a more bullish scenario for the host government leaders, in which the country's hydrocarbon potential far exceeds even the most optimistic projections. On its way to becoming an important source of crude entering world trade, the government seeks to mimic the success of other oil exporters in capturing a larger share of the economic rents generated by the petroleum sector. In addition to equity participation in new projects (our model is left unaffected by this change in policy), the goverment introduces increasingly onerous rates of royalty assessment and taxation. With royalties raised to 16.67 percent, the government begins to increase the tax bit. With taxes at 60 percent, net income slides to less than 30 percent of the original projection; at 80 percent, net income under scenario G is barely 15 percent of that promised under the initial terms operational under case A.

The OPEC-sanctioned rates, however, have not yet been realized. Scenario H, in which the government imposes the OPEC target rates of a 20 percent royalty and an 85 percent income tax, pushes the company into the red. The royalty payment reduces net revenues to the point at which they are insufficient to cover operating costs. The project encounters losses. At this extreme, the oil firm has been *de facto* expropriated.

The hypothetical example reveals the potential impact of one type of political risk event on the net income derived from an investment. Although our scenario in table 7.1 leans toward the extreme, it is apparent that changes in the fiscal regime can alter the company's income sheet. Projections of future income based on the initial case would have resulted in a gross overvaluation of the worth of an investment. If anticipated fiscal changes had been incorporated into the ledger sheet projections, the firm would have been better able to assess the value of a project. Perhaps the company would have made the investment anyway, confident that it would recover its outlay and earn a handsome return before a particularly onerous fiscal regime took effect. This, however, would have been a well-reasoned decision based on a forecast of how economic, geological, and political variables will affect future profitability, rather than a naive projection of today into the future.

As discussed in chapter 6, using an arbitrary discount rate or levying a risk premium on investments in risky countries fails to address the subtle problems of political risk analysis. In fact, these procedures invite abuse in misusing political risk as a variable in decision making by prejudicially favoring some investments over others. Nonetheless, political risk discount rates for classifications of countries frequently are used in evaluating investment agreements, although such approaches are more common in the nonextractive sectors.

Table 7.2 presents a simplified hypothetical cash flow showing the effect of different discount rates on the profitability of an investment. Discounting for the time value of money (scenario A), the $25 million cash flow annually for ten years yields a net present value of $45.01 million. The initial investment is recovered in 4.4 years. To account for differing political risks, the discount rates have been adjusted to apply to investments in three country groupings. Let us assume that the Discount Oil company uses the 15 percent rate (A) on domestic investments. Investments abroad are discounted according to the risk class in which a country is placed: 20 percent (B) is used for investments in politically safe countries, 25 percent (C) applies to moderately safe states, 30 percent (D) is the rate for moderately risky countries, and projects in risky nations are not considered. In effect, this method levies a political risk premium of 5 percent, 10 percent, and 15 percent on countries in categories B, C, and D respectively.

As the table indicates, the different discount rates have a substantial impact on the cash flow projections. Compared to the domestic scenario (A), the same project in countries in class B has a net present value equal to 41 percent of that in class A, and the payback period is extended by just over one year. In moderately safe countries (C), the net present value represents 31 percent and only 13 percent of that under classes B and A respectively. The payback period has been further stretched. The same $25 million for ten years in a country under category D yields a negative net present value, indicating that the project would be rejected.

Rather than applying an arbitrary discount rate to gross categories, political risk can be factored into cash flow projections on a country- and investment-specific

Table 7.2   The Effect of Higher Discount Rates on a Hypothetical Cash Flow of
$25 Million Annually for Ten Years
($000)

	Discount Factors[a]			
	A *15%*	*B* *20%*	*C* *25%*	*D* *30%*
Preproduction investment	(75,000)	(75,000)	(75,000)	(75,000)
Present value	120,010	97,635	80,795	67,900
Net present value	45,010	18,635	5,795	(7,100)
Payback period (years)	4.4	5.5	7.7	—

[a] Annual continuous compounding rate.

basis. This can help decision makers evaluate an estimated cash flow that has been developed in light of political risks. Building on the original scenario in table 7.2, table 7.3 presents a synopsis of a hypothetical cash flow modified to reflect four sets of assumptions about the risk of expropriation. The annual time-discounted cash flows have been multiplied by the probabilities (political risk adjustment factors) of the project not being expropriated to determine the most probable estimated cash flow for each year.

Higher risks of expropriation reduce the expected value of future cash flows. Assuming no risk of expropriation in the first five years of operations and a 50 percent probability during the second half of the project, investment A has a net present value of 58 percent of an investment without expropriation risk. Although the net present value is smaller in investment A, the payback period remains unchanged. This is because the projected cash flows for the first five years are the same in both instances. Because of the time value of money, maximizing returns in the early years of a project may compensate for future risks. If the risk of expropriation were 50 percent for each year, the investment would be rejected (because of a negative net present value). Realizing full returns in the first five years is sufficient in our example, however, to insure a positive net present value. In contrast to adjusting the discount rate to 20 percent (see table 7.2), investment A yields a larger net present value and shorter payback period and presents a more realistic picture of annual cash flows.

Investments B and C are adjusted to reflect more serious risks of expropriation than case A. Although the net present value slips by almost 25 percent from case A to case B, the payback period is largely unaffected. The value of investment C is severely reduced in terms of net present value and the length of time necessary to recover the initial investment. Even in case C, however, where the risk of expropriation grows by 10 percent annually, the net present value is positive and is almost as great as that realized by a straight 25 percent discount rate (see table 7.2). It is only under the extremely risky scenario presented in case D that our hypothetical cash flow returns a negative net present value because of the likelihood of expropriation. Even in this case, if the cash flow in year 2 remained unaffected by expropriation risk, the net present value of investment D would have been positive.

The hypothetical probabilities of expropriation in table 7.3 represent a series of most-likely scenarios. In effect, these probabilities are based on a range of potential outcomes. Treating the best estimate of the distribution of a range of probabilities simplifies the problem of computing cash flows along a spectrum of estimates. Rather than simply dealing with the most-likely scenario, however, a cash flow can be constructed along a range of probabilities or a series of probabilities, including a most-likely, a pessimistic, and an optimistic scenario. This might provide additional insight into the value of an investment. In particular, projections based on optimistic and pessimistic cases will yield a band of values within which the most-likely scenario and actual cash flows almost certainly will fall.

Table 7.3 Hypothetical Cash Flow of $25 Million Annually for Ten Years Adjusted for Time and Expropriation Risk ($000)

	Time Discount (15%)[a]	Investment A	Investment B	Investment C	Investment D
Preproduction investment	(75,000)	(75,000)	(75,000)	(75,000)	(75,000)
Present value	120,010	100,758.8	94,753	80,322.3	73,605.8
Net present value	45,010	25,758.8	19,753	5,322.3	(1,394.2)
Payback period (years)	4.4	4.4	4.6	6.7	—
Annual likelihood of *not* being expropriated:					
Year 1		1.0	1.0	1.0	1.0
Year 2		1.0	1.0	0.9	0.75
Year 3		1.0	1.0	0.8	0.75
Year 4		1.0	0.8	0.7	0.6
Year 5		1.0	0.8	0.6	0.6
Year 6		0.5	0.8	0.5	0.4
Year 7		0.5	0.4	0.4	0.4
Year 8		0.5	0.4	0.3	0.2
Year 9		0.5	0.4	0.2	0.2
Year 10		0.5	0.2	0.1	0.1

(a) Annual continuous compounding rate.

165

Expropriation risk is unique in that it is an all-or-nothing situation. A project is either expropriated or it is not expropriated. A 75 percent likelihood of not being expropriated in year n means that there is a 75 percent probability of the anticipated cash flow being realized and a 25 percent chance of a cash flow of zero. Multiplying the expected cash flow by the probability of not being expropriated (see table 7.3) is the equivalent of taking the weighted arithmetic mean of the likelihood of an uninterrupted cash flow (no expropriation) and no cash flow (expropriation).

Other risk events are not so clear cut as expropriation in their impact on cash flows. Repatriation restrictions, for example, do not imply simple alternatives of free remittance or no remittance abroad. Rather, repatriation may be limited to some fixed amount or percentage of earnings. Instead of directly limiting the flow of profits abroad, the host government may prefer to set reinvestment targets. The effect—curtailed movement of corporate earnings—is the same.

A 10 percent probability of restrictions on the repatriation of profits can be operationalized in a number of ways. Assuming a constant cash flow, the 10 percent likelihood of restrictions may mean that, for every income period, the firm can repatriate 90 percent of its earnings, that the company can remit abroad 100 percent of the first nine-tenths of income and afterward cannot send profits home, or that, for every ten days of operations, nine days' worth of earnings can be repatriated while one day's worth cannot. There is an infinite number of ways in which a 10 percent likelihood of repatriation restrictions can be operationalized. Conceptually, however, the 10 percent probability of repatriation restrictions means an estimate that the company will realize a cash flow abroad equal to 90 percent of what would have been the case without restrictions.

Table 7.4 presents an expanded hypothetical cash flow for a project, taking into account a number of political risks. In each instance, the current dollar cash flows have been multiplied by the discount and adjustment factors in the second part of the table. The time value discount factors are based on the same 15 percent continuous annual compounding interest rate used previously in this section, while the risk event adjustment factors are hypothetical. Each risk event is treated discretely, with no account taken of multiple risk events.

The risk-adjusted cash flows have been developed cumulatively across each row; that is, the time-discounted values have been multiplied by the production cut adjustment factors and the result was then multiplied by the adjustment factors for restrictions on sales abroad, and the like. The column headed "Expropriation," therefore, reflects the cash flow adjusted for all five categories of risk, "Repatriation Restrictions" incorporates the first four categories, and so on.

Without incorporating political risk analysis into the cash flow, the proposed project has a net present value of $41.13 million, promises a recovery of the initial investment in 7.4 years, and yields a 21.5 percent internal rate of return on capital. The host is not likely to impose production cuts on the firm. In years 3 and 4, the host government is expected to restrict liftings for conservation reasons such that only 95 percent of the anticipated cash flow is realized. Otherwise, the country's shortage of foreign exchange and need to satisfy growing domestic consumption make it likely that the firm's production schedule will be fulfilled. The slight setbacks in years 3 and 4 have a minor impact on the cash flow. Net present value falls by less than 4 percent, the payback period is extended by 0.10 year, and the internal rate of return declines 0.3 percent.

Growing domestic consumption increases the chances for restrictions on sales abroad (independent of changes in production levels). As sales abroad decline in years 4 through 8, new explorations and projects are undertaken. As more production is brought online, sales abroad increase, stabilizing at 90 percent of anticipated earlier levels for years 12 through 15. The cumulative risks thus far (production cuts and restrictions on foreign sales) mean a drop of almost one-quarter in the project's net present value. Similarly, the payback period is extended and the internal rate of return falls.

Moving across the columns, incorporating additional political risk factors results in reductions in the adjusted cash flow for the project. The tax regime becomes increasingly costly after year 5. Between 5 and 10 percent of corporate earnings are blocked from repatriation to ease the foreign exchange problem. Expropriation, though still not likely, becomes more probable as the project matures. The net present value and internal rate of return decline, while the payback period becomes longer.

The relatively mild nature of the risks in years 1 through 5—when cash flow is worth the most in time-discounted dollar terms—helps make the project profitable even when all five categories of risk are taken into consideration. More than 88 percent of the initially projected cash flow for the first five years is realized after the political risk adjustments are made. More than 63 percent of the investment and almost 60 percent of the risk-adjusted net present value of the project are achieved by year 5. The bottom line, so to speak, is the last column. Having accounted for all the categories of risk included in table 7.4, the project still has a positive net present value and should be accepted (assuming that there are no other projects with a greater net present value).

As was mentioned earlier, we have treated only the most likely values. A more optimistic scenario in table 7.4 would result in a more-favorable cash flow. Conversely, worst-case assumptions would reveal a less-profitable investment proposition. Table 7.5 incorporates a worst-case expropriation scenario in place of the expropriation column included in table 7.4. The adjustment factors—revealing the probability of not being expropriated—have been reduced substantially, indicating a far greater likelihood of expropriation commencing with the second year of operations. The accelerated expropriation schedule assumed in table 7.5 results in a cash flow that merits rejection of the project; the investment carries a negative net present value and the internal rate of return is 14.5 percent, lower than the market rate of interest.

Table 7.4 Hypothetical Cash Flow from Petroleum or Natural Gas Investment Adjusted for Time and Political Risks ($000)

	Traditional Cash Flow		Political Risk-Adjusted Cash Flow for Five Risk Events (cumulative)				
	Cash Flow (Current $)	Time-Discounted Cash Flow (15%)	Production Cuts	Restrictions on Sales Abroad	Costly Fiscal Changes	Repatriation Restrictions	Expropriation
Preproduction investment	(100,000)	(100,000)	(100,000)	(100,000)	(100,000)	(100,000)	(100,000)
Year 1	15,000	12,911	12,911	12,911	12,911	12,911	12,911
Year 2	20,000	14,816	14,816	14,816	14,816	14,075	13,371
Year 3	25,000	15,940	15,143	15,143	15,143	14,386	13,667
Year 4	25,000	13,720	13,034	12,382	12,382	11,763	11,175
Year 5	30,000	14,172	14,172	13,463	13,463	12,790	12,151
Year 6	30,000	12,198	12,198	10,978	9,880	9,386	8,447
Year 7	35,000	12,247	12,247	11,022	9,920	8,928	8,035
Year 8	35,000	10,542	10,542	8,961	8,065	7,258	6,532
Year 9	35,000	9,076	9,076	8,168	6,943	6,596	5,936
Year 10	35,000	7,809	7,809	7,418	6,305	5,990	5,092
Year 11	30,000	5,763	5,763	5,475	4,654	4,421	3,758
Year 12	30,000	4,959	4,959	4,463	3,794	3,604	3,063
Year 13	25,000	3,558	3,558	3,202	2,401	2,281	1,824
Year 14	15,000	1,838	1,838	1,654	1,240	1,116	837
Year 15	15,000	1,581	1,581	1,423	925	833	625
Present value		141,130	139,647	131,479	122,842	116,338	107,424
Net present value (NPV)		41,130	39,647	31,479	22,842	16,338	7,424
Payback period (years)		7.4	7.5	8.0	8.5	9.3	10.7
Internal rate of return (0.1%)	21.5		21.2	20.0	19.0	18.0	16.5

Table 7.4 (*continued*)

Discount and Adjustment Factors

Year	Time Value [a]	Production Cuts [b]	Restrictions on Sales Abroad [b]	Costly Fiscal Changes [b]	Repatriation Restrictions [b]	Expropriation [b]
1	.8607	1.00	1.00	1.00	1.00	1.00
2	.7408	1.00	1.00	1.00	0.95	0.95
3	.6376	0.95	1.00	1.00	0.95	0.95
4	.5488	0.95	0.95	1.00	0.95	0.95
5	.4724	1.00	0.95	1.00	0.95	0.95
6	.4066	1.00	0.90	0.90	0.95	0.90
7	.3499	1.00	0.90	0.90	0.90	0.90
8	.3012	1.00	0.85	0.90	0.90	0.90
9	.2593	1.00	0.90	0.85	0.95	0.90
10	.2231	1.00	0.95	0.85	0.95	0.85
11	.1921	1.00	0.95	0.85	0.95	0.85
12	.1653	1.00	0.90	0.85	0.95	0.85
13	.1423	1.00	0.90	0.75	0.95	0.80
14	.1225	1.00	0.90	0.75	0.90	0.75
15	.1054	1.00	0.90	0.65	0.90	0.75

[a] 15 percent interest, continuous compounding.

[b] Each factor is computed independently, although the effects have been calculated as cumulative. The factors represent the most probable impact on the projected cash flow. Conceptually, a 0.95 factor for production cuts in years 3 and 4 indicates the likelihood that production cuts for those years will limit the firm's cash flow to 95 percent of the expected amount. Restrictions on sales abroad (independent of cuts in production), costly fiscal changes, and repatriation restrictions are conceptualized in the same manner. Expropriation is conceptually different, in that it implies an all-or-nothing scenario. In this instance, the factors are the probabilities of not being expropriated.

Table 7.5   Worst-Case Expropriation Scenario Substituted for
"Expropriation" Column in Table 7.4
($000)

	Worst-Case Expropriation	Adjustment Factor
Preproduction investment	(100,000)	
Year 1	12,911	1.00
Year 2	12,668	0.90
Year 3	12,947	0.90
Year 4	8,822	0.75
Year 5	9,593	0.75
Year 6	7,040	0.75
Year 7	4,464	0.50
Year 8	3,629	0.50
Year 9	3,298	0.50
Year 10	2,396	0.40
Year 11	1,768	0.40
Year 12	1,442	0.40
Year 13	752	0.33
Year 14	368	0.33
Year 15	275	0.33
Present value	82,373	
Net present value	(17,627)	
Payback period (years)	—	
Internal rate of return (0.1%)	14.5	

Table 7.4 presents a way for decision makers to evaluate an investment both in strictly financial terms and in accordance with the most probable political developments. Keeping the financial projection separate from political variables—something that is not done by adjusting discount rates (see table 7.2)—is important in assessing an investment in market and geological terms and then in terms of political developments. This enables decision makers to identify the impact political variables have on cash flows.

Although our computations were made in a cumulative manner, the figures can be disaggregated to show the adjusted cash flow for each risk category or group of categories. Multiplying the time-discounted cash flow by the column of adjustment factors for each category of political risk reveals the impact each risk presents individually. The net present value of the project based on separate computation of political risk adjusted cash flows is as follows:

Production cuts	NPV	=	$39,647 million
Sales restrictions	NPV	=	$32,928 million
Fiscal changes	NPV	=	$31,590 million
Repatriation restrictions	NPV	=	$33,407 million
Expropriation	NPV	=	$29,445 million

Individually computed, expropriation risks present the most serious threat to the proposed investment, whereas the likelihood of production cutbacks is the least significant political problem with which the firm may be confronted. Shown in this fashion, the separate political risks may be better comprehended by the firm. This also helps direct decision makers and political risk managers to the political issues that are the most important in insuring the future profitability of an investment. Similarly, computations can be based on a group of political risk events to reflect the combined impact of any combination of political risks on cash flows.

The political risk adjustment factors in table 7.4 are hypothetical, however. As is the problem with some of the political risk forecasting techniques surveyed in chapter 6, such as the decision tree, a model for constructing political risk-adjusted cash flows does not give us any direction in terms of selecting adjustment factors. The adjusted cash flow approach, therefore, must be combined with some means of estimating the probability of events. This can be done by means of the Prince system, the Delphi method, or any of the other approaches.

Assuming that reasonably accurate measurable probabilities of political risk events are possible in terms of their likely impact on a project, the political risk-adjusted cash flow is a way for firms to see clearly the effect that political developments are likely to have on an investment. Again, in addition to the most-likely case, best- and worst-case scenarios can be developed to show the range of probable cash flows. Although the overall risk-adjusted cash flow should be the basis for final evaluation of a project, separate risk events and groups of events can be measured for their impact on profitability. This will point toward the potential political risk areas that are most important to the firm and may prove useful in efforts to manage political risks.

Obviously, the political risk-adjusted cash flow can be no better than the data and methodology used in computing the adjustment factors. The same, however, applies equally to the traditional cash flow estimated on geological and economic inputs. The "garbage in, garbage out" problem may be more serious in political risk evaluation than in geological and economic measurements and calculations, but geological and economic data are also subject to multiple interpretations and ambiguities. Although political variables may be more diffuse and softer than geological and economic parameters, they are still estimatable and can be incorporated into company cash flow projections.

Although political risk adjustment factors may lack accuracy, the risk-adjusted cash flow should present a more accurate and realistic profile of the expected cash flow from an investment. It must be borne in mind that not adjusting for political risks is functionally the same as using political risk adjustment factors of 1.0—that is, the assumption that political risk events will limit the firm's annual cash flow to 100 percent of what was expected. Seen in this way, it is apparent that *not* adjusting cash flows for political risks is equivalent to adjusting for a risk-free environment. Although risk-adjusted cash flows are difficult to construct and have measurement problems, the alternative promises an unrealistic projection of current conditions into the future.

Managing Political Risk

Political risk analysis is most commonly employed in the early stages of an investment decision. Before it makes any sizable investment commitment is the most logical time for a company to assess the investment climate and political risk. In the petroleum and natural gas industry, as in the minerals sector on the whole, large investments are necessary for exploration. Exploration may predate drilling by three to five years, after which another six months to two years may elapse before production starts. Typically, therefore, the company assesses the political environment prior to undertaking exploration, which may be four to seven years in advance of commercial production.

Although the lead time is not so long as it is in the hard rock minerals industries, it is longer than for most other types of investments. The risks in exploration—the probability of *not* finding commercially exploitable fields—are far greater in the extractive sector than parallel risks are in other enterprises. It is because of the lead time and exploration risks that companies prefer to negotiate exploration/production agreements prior to exploration. It is at some time before the conclusion of an investment agreement that the company must analyze the political risks and make a decision about whether to pursue a project that may first come onstream in five years or more and then may be scheduled to run anywhere from one to three decades.

In most instances, political evaluation begins with the firm's announcement that it is considering a project abroad. The internal stimulus to analyze the political situation may be even more specific; in response to a high-level directive to look into the prospects for investing in a given country, a political risk evaluation may be initiated. Companies rarely assess political risk in advance of an expression of corporate interest in a potential project. In many instances, firms do not even collect basic information about nations in which they do not have and are not considering a particular investment.

After an investment has been made, political risk forecasts are frequently employed with respect to questions about the reinvestment of earnings. In many instances, however, the political risk assessment function is seen as limited to the initial investment period, and political concerns are ignored or treated implicitly when it comes to questions of reinvestment. This indicates a narrow conceptualization of political risk management as well as a universal application of the laws of thermodynamics: a project in motion will remain in motion until met by an opposite and equal force. By that time, however, the damage may be done. There is an inherent bias toward continuing and expanding operations that are in progress. Ongoing operations often are invoked as evidence that the country's political environment is favorable and that there is no need to reassess the political scene. This type of circular deductive reasoning, in which the conclusion is made that a country is politically conducive to reinvestment because an initial investment was made, has sometimes been invoked to short-circuit any inquiry into political risk.

Another area in which political risk is considered important by a number of firms is strategic planning. By definition, strategic planning is concerned with the future course the company will take in the medium to long term—say, five to ten years and beyond. In plotting a course into the future, the firm must make certain assumptions about conditions years hence. This necessarily entails forecasts of political conditions and changes as well as economic and market developments. In many respects, strategic planning lends itself strongly to political risk analysis, as both have lengthy time horizons and are concerned with predicting changes and their effects on the firm.

The concept of political risk management reaches beyond invoking risk analyses only in response to immediate needs with respect to an investment or reinvestment. Rather, as mentioned earlier in this chapter, the management of political risk includes forecasting as only one step in the continuous process of keeping aware of how political conditions are changing, the ramifications for the firm, and the policy responses of the company:

> *Risk management entails identifying risks, assigning a value to them, anticipating losses, and making objective decisions about what steps to take before losses occur so that they have the least impact on the operations of the enterprise. It also includes a loss control program in order to help prevent or reduce the incidence or severity of losses.* [5]

Integrating political forecasts into strategic planning is moving in the direction of political risk management—the balancing, controlling, and executing of company goals and interests in light of a changing political environment.

Rather than maintaining a political risk management function, the tendency is to initiate risk analyses in response to a particular situation. Forecasts on demand are responses to specific needs by decision makers. As previously noted, the demand typically comes upon consideration of a new investment or, to a lesser extent, consideration of reinvestment. Because of the failure to treat risk management on an ongoing basis, which includes monitoring of changes and updating of analyses, firms often fail to attempt to manage problems until they arise. "In coping with political risk, most corporations are engaged in response rather than initiation,"[6] reacting to changes that are already upon them, rather than predicting and preparing for future developments. Responses after the fact severely constrain the firm's flexibility in dealing with circumstances and indicate a failure in keeping pace with changing conditions.

Once an investment has been made, political variables all too often are ignored until something dramatic happens or the company is confronted with developments that adversely affect its interests. Managerial and operating decisions rarely indicate any concern with political risk. Risk forecasts are infrequently used with respect to company policy regarding divestment, repatriation, and other exchange operations and are even less common in day-to-day operating decisions.[7]

Given the frequency of occurrence and the magnitude of political events surrounding international energy issues—particularly oil and gas—one would expect a full-fledged effort to develop a political risk management function in transnational energy companies. Despite general recognition of political problems, however, most firms have not moved beyond the stage of forecasting political developments. As noted earlier, political risk analysis tends to be separated from the decision process. Monitoring is typically conducted on an irregular basis, with countries in which the company has an investment being reviewed at intervals ranging from six months to two or three years. If something happens in the interim that may affect the firm, a review will be undertaken at that time. Other nations usually are ignored, unless corporate management suggests a potential future interest.

Monitoring, moreover, which may or may not entail a complete update of the initial political risk analysis, almost never is done in light of possible changes in company policy in an effort to manage future risks. To a large extent, this reflects the organizational logjam in communications between risk analysts and decision makers. As difficult as it is to get political risk analyses integrated into the initial decision-making process with respect to an investment abroad, it is a far more difficult task to maintain a political risk management function (including revision of earlier forecasts) that has regular input to decision makers and managers with respect to either general policy or operating decisions. Lacking such feedback on political developments, most firms initiate efforts to manage political risks only *after* risk events have occurred and the firm is suffering the consequences or is imminently confronted with potential losses.

Oil and gas companies do not have great latitude in managing political risks. Many of the techniques used by firms in other sectors—such as specializing production so that any given country hosts a plant that manufactures only a component of the finished product,[8] producing for a specialty market in which the TNC is the sole (monopsonist) buyer of the product, or literally transplanting facilities to different locations and jurisdictions—are not appropriate or have only minor applications to firms producing petroleum and natural gas. The extractive industries have no choice but to extract (or not extract) raw materials where they are found in the ground. Production facilities cannot be disassembled and moved elsewhere. The OPEC countries, moreover, have broken the companies' previous tight control on transportation, refining, and marketing, further restricting the firms' ability to insulate themselves from political risk.

This does not mean that firms in the extractive sectors have no opportunities to try to minimize their political risks. Although their scope of risk management activities may be more circumscribed, oil and gas firms can take an active role in attempting to control their exposure to political risks. This is the purpose of risk management.

Strategies and Tactics

Modern transnational firms have largely overcome the traditional business perspective of the marketplace as a simple economic arena of supply and demand. Political forces are recognized as important determinants of foreign investment projects and world trade. It was this recognition that prompted companies to pay closer attention to political variables and to move toward political risk analysis as a standard procedure in evaluating the merits of a project abroad.

Moving beyond passive acceptance (or rejection) of political risks, companies have begun to play a more active role in asserting their interests by trying to manage their exposure to political risks. This is part of the process of developing a political risk management function. Having come to accept the political nature of foreign investment, companies are beginning to realize that they cannot be spectators with respect to how their interests are affected by political developments.

The simplest risk management tactic is avoidance. Analysis of a country's political environment may point to excessive risks or may compare unfavorably with alternative investment opportunities. In such an instance, corporate decision makers would be expected to decide against the investment, thus avoiding the risks. When decisions are based on a thorough examination of all relevant variables—including political ones—avoidance needs to be seen as one approach to the problems of managing political risks. Reflexive avoidance of projects because of superficial impressions or faulty understanding, on the other hand, is not part of a strategy of risk management. Even though the end result with respect to the particular prospective project may be the same, reflexive avoidance flies in the face of efforts at serious consideration of political risks and may be described more aptly as risk *non*management.

Domestically, U.S. firms have a lengthy history of political activity aimed at shaping their futures. Mass publicity campaigns aimed at the general populace and lobbying of government officials are part of the standard fare of corporate activity. This is seen as a legitimate expression of corporate interests and the competition for influence in a pluralistic political system. Similarly, in developed countries abroad, U.S. firms regularly participate in the political process, though to a much lesser extent than at home.

There are a number of sensitive issues surrounding political activities of companies abroad. The host country may not see the foreign firm as having *any* legitimate role to play in the domestic political process. As a foreigner, the TNC is expected to behave in accordance with local standards and laws and may be discouraged from trying to influence the course of national political events. Publicity or lobbying efforts may be seen as unwarranted foreign interference—an illegitimate expression of outside influence that threatens the state's sovereignty.

Developing countries, ever guarding their hard-earned independence and protecting their national sovereignty, are particularly likely to reject foreign corporate efforts at influencing domestic political events. The companies frequently are seen as foreign policy projections of the home government. Cries of neocolonialism and neoimperialism are almost certain responses to company efforts at participating in national politics in Third World countries. Direct attempts at influencing political leaders or the relevant political audience are likely to be seen as illegitimate and may result in political backlash against the firm. Political visibility can prove to be a liability for foreign companies. Given the size of transnational petroleum and natural gas firms and the resources they command compared to the size and resources of many developing states, it is not surprising that host governments in the Third World strongly object to political activities sponsored by foreign firms.

OECD nations tend to be more mutually tolerant of political activities undertaken by TNCs. These countries usually are more secure in the integrity and durability of their political processes and more committed to democracy. They also tend to be the home base for their own transnational firms that have interests to protect abroad. Even in OECD states, however, governments and the people have at times grown intolerant of the political activities of foreign firms. Concerns about the national interest, a growing sense of resource nationalism, and latent political nationalism often circumscribe the activities of foreign extractive sector TNCs, even in the Western countries.

Companies have an obvious and understandable concern about protecting their interests. They may not, however, be afforded a legitimate role in the national political process, or they may have only limited leeway. Typically, therefore, TNCs must restrict themselves to managing political risks and protecting their interest in the face of risk events rather than trying to influence the political fate of the country.

In many respects, this is similar to treating the symptoms rather than the cause of the problem and restricts the options available to companies that are eager to manage their exposure to political risks abroad.

Different types of political risks are more amenable to management than others. Most risk management strategies focus on ownership risks (see chapter 1) and the protection of a company's equity position in an investment. Protecting against transfer and operational risks is less often considered, perhaps because of the sense of reduced options available to the company. Administrative/statutory risks present especially difficult problems, as they are closely linked to the legislative and bureaucratic processes of the host country. Finally, the ability of transnational oil and gas firms to manage the contractual risks to which they are exposed may be more directly related to market supply and demand forces, as opposed to corporate policies, than is the case with the other classes of risks.

Most risk management strategies and tactics are general in their focus, rather than being detailed to address a particular type of risk. The maintenance of an amiable and workable host-firm relationship, for example, may be the most important aspect of a risk management strategy. This is an ambiguous concept, however, which gives little direction in terms of tactics to be employed and does not seem to be causally connected with any of the types of political risks we have identified. Political risk insurance, on the other hand, tends to be risk-specific, addressing such particular concerns as expropriation (ownership risks) or restrictions on remittances (transfer risks).

There is no single tactic or strategy that is certain to protect a company from political losses. The risk management strategy must be tailored to the particulars of an investment. There is an array of policies a company can implement to discourage, slow down, or raise the costs to the host country of costly changes in the investment regime. Firms should pursue as many avenues of risk reduction as are available and prudent, as there is no simple solution to the problems of political risk. Specific tactics, moreover, often are a two-way street: what helps reduce risk in one instance may aggravate the situation in another. This points to the need for firms to be flexible and pragmatic, managing risks in a fluid environment without losing sight of circumstances.

The enlistment of home government support can be such a two-edged sword. Confronted with politically based threats to their investments, companies often appeal to their home government for backing. This is the standard response to the possibility of expropriation in particular. The age of gunboat diplomacy, however, is over. In its stead, companies turn to their home governments to apply political and economic pressures against recalcitrant host states.

The United States has a lengthy legislative history of economic sanctions against countries that expropriate U.S. corporate assets without proper compensation. The 1959 Johnson-Bridges Amendment required the suspension of aid to an expropriating country within six months of an uncompensated expropriation, with the proviso that the president could waive application to protect the national interest. As a result of dissatisfaction with the presidential prerogative permitted by the 1959 law, the Hickenlooper Amendment was passed in 1962. This required the termination of aid to any country that failed to pay proper compensation, without giving the president the option to decide otherwise. The following year the amendment was broadened to include investment contracts that were unilaterally ended by a foreign country. (Ten years later, the Hickenlooper Amendment was made voluntary, depending on presidential initiative.) In 1972, the Gonzalez Amendment broadened the scope of sanctions, providing for U.S. refusals on votes for loans by international financial institutions to countries that refuse compensation. In addition, the United States has passed scores of amendments on other laws, such as the 1962 Sugar Act, providing for sanctions against nations that seize U.S. corporate assets without compensation.

There is scant evidence that such sanctions have been effective. Certainly the threat of sanctions has not deterred the onslaught of nationalizations and expropriations in hydrocarbons, as well as in other sectors. In all fairness, however, it is impossible to know if any particular seizures were discouraged or delayed because of potential sanction. It is likely, moreover, that the possibility of sanctions encouraged countries to resolve the compensation issue *after expropriation* more quickly than otherwise would have been the case.[9] Although this may be after the fact, prompt and adequate compensation is of vital importance to a company that has had its foreign assets expropriated.

Enlisting government support has not always worked. The U.S. government traditionally has been reluctant to impose sanctions. Ten years after its enactment, the Hickenlooper Amendment was applied only once, and that was to Sri Lanka (Ceylon). Fearful of the potential political ramifications—that the sanctioned country, as in the Cuban experience, might turn to the Soviet Union—and the chance of backlash against other U.S. investments in the country involved, the government has frequently threatened but rarely imposed sanctions. Because of such political and economic concerns, and despite Exxon's insistence, the Nixon administration did not invoke the Hickenlooper Amendment against Peru in response to the 1969 expropriation of the International Petroleum Company. The Peruvian government made no effort to veil its intention to retaliate against American companies if the United States imposed sanctions. To the extent that sanctions have been levied, especially in Latin America, one analyst concludes: "If anything, formal aid sanctions have only strengthened nationalist hostility toward foreign investment."[10]

Sanctions, it seems, have not prevented expropriations. The government's reluctance to resort to sanctions further undermines the credibility of the economic threat. Waving a stick at Sri Lanka in the early 1960s is significantly different from brandishing threats at the OPEC countries in the mid-1970s. Expropriation, moreover, is only one type of political problem for TNCs, albeit the most extreme.

There has been some effort to expand the protections to creeping expropriation and such risk events as unilateral nullification of contracts or imposition of discriminatory taxation, but this is unlikely to produce any better results than have been experienced with respect to uncompensated expropriations. Many other risk events—such as restrictions on foreign exchange and repatriations or production cutbacks—are not addressed under U.S. legislation.

Calling on the backing of the home government may enrage host government leaders. Third World regimes often see transnational firms as extensions of their home governments—a suspicion that would be confirmed by such a call for help. After trying to establish an independent transnational identity for themselves, distinct from national affiliations, falling back on the support of their home governments will undercut the claim of TNCs that they are nonpolitical industrial enterprises. Moreover, host leaders can reply with more than a little validity, that the U.S. government has no standing or role to play in what can be seen as a domestic political issue. Calling for support from home is inviting foreign interference in the domestic life of the host country.

A last problem, which typically is unmentioned, is that the home governments may refuse to support the company's claim. As in the Peruvian case, the government may have different priorities than the firm and will not lend its full weight in defense of the firm's position. The growing differences between the interests of U.S.-based extractive firms and the government (see chapter 2) make it increasingly likely that the government will not be willing to throw its complete support behind the affected company. National security interest, in particular, may be afforded greater consideration than the particular economic losses confronting one or a number of corporations.

In addition to mobilizing the support of the home government, an act that typically is a response to an imminent problem, the firm can try to develop political support in the host country. Wooing host government political leaders can be very tricky and risky. A change in government may mean a loss of patronage. To the extent that the firm is identified with a particular regime, it may have problems in the event that an opposition group comes to power.

In any political system, political favor and backing is often bought by an exchange of supports. Corporations are not unfamiliar with the tactics of making contributions to or campaigning for political parties and individual candidates. Although this may be the norm at home, however, a foreign firm that contributes to campaign or political funds abroad is a likely target for charges of meddling and interfering in domestic politics. At times, firms resort to under-the-table payments or bribes to gain political influence. This may breed further government corruption, a malaise that often undermines the host government's legitimacy and creates political instability. Corruption of this sort, which is not uncommon in the developing world, can implicate foreign firms in the unsavory policies of an unpopular government, tainting it by identification with an odious regime that is destined to be overturned. The loyalties of host government officials that are bought with bribes, the giving of which is illegal under U.S. law but widely tolerated elsewhere, are fickle and are not assurance of support.

In most instances, prudent judgment would reason against becoming involved in national partisan politics abroad. Broader, more lasting political support might be better built by "identify[ing] the firm with national aspirations and objectives."[11] This type of support should be nurtured both from above, among national elites and decision makers, and from below, among the citizenry at large. Rather than aligning itself with a particular party or political figure, the company might be wiser to side with the national interest and welfare of the people.

To diversify their base of support internationally, firms often multilateralize the financing of projects. By spreading the equity in a project over a number of companies from different countries, the legal, political, and economic obstacles to unilateral changes in agreements or expropriation are increased. In the event that politically induced losses are inflicted on such a consortium, the host government will be confronted with the opposition of a condominium of firms and home governments. As is typical of many risk management options, spreading the equity among firms from different countries raises the costs to the host government of politically interfering in a project's operations.

Another means of diversifying the financing of an investment is to spread the debt position among banks from different countries and multilateral financial institutions. Rather than using corporate funds, the oil company raises debt capital from banks in, perhaps, the United States, Japan, and West Germany. The separate project is legally constructed to succeed or fail in its own right, without obligation on the part of the mother company. The ability of the banks to be repaid depends on the profitability of the project. Repayment typically comes from future sales, giving the banks an ongoing interest in operations. Production cutbacks or restrictions on sales will have an obvious adverse effect on the banks' position, as would restrictions on remittances abroad. Although the banks may not have a direct equity position in the investment, they have a tangible stake in the outcome. Reducing the profitability of a project will alienate the banks that have bankrolled the investment as well as their home governments. The host country's ability to borrow abroad will suffer immensely in the face of any actions that mean losses for the transnational banks.

Involving the World Bank (formally known as the International Bank for Reconstruction and Development) or any of the other multilateral global or regional banking institutions gives a company an additional measure of protection. It is unlikely that a host government could afford—either politically or economically—to cause losses to the Bank because of interference in a Bank cosponsored investment. Typically, the Bank contributes a relatively small share to a project, expecting that private financing will make up the difference. The Bank's oil and gas program encourages, and at times requires, private cofinancing, providing "a form of 'ex-propriation insurance' to private co-financing partners."[12]

The Bank does not see itself as a substitute for private financing. In many respects, it acts as a lightning rod for private capital. Bank officials estimate that a dollar's worth of Bank loans or credits stimulates two dollars of private capital invest-ment during the exploration period and three dollars during the development stage.[13] Moreover, like the other multilateral financial institutions, such as the International Monetary Fund and the Inter-American Development Bank, the World Bank largely adheres to traditional principles of financing. Loans are not extended to countries that are considered bad risks; nor is the Bank willing to lend to states that have outstanding investment disputes—particularly about uncompensated expropria-tions—that the host government is not negotiating in good faith. To a large extent, this reflects the voice exercised by the United States and other developed countries in setting the Bank's policies.

Yet another method oil companies have used to diversify the financing of an in-vestment is factoring, whereby the firm concludes long-term sales agreements with buyers and sells the contracts to financial intermediaries that collect on the contracts.[14] Although this is an expensive practice, it shifts the transfer risks of losses caused by cuts in liftings or sales abroad onto the shoulders of the financial mid-dleman. As in diversifying the direct equity or capital debt of an investment, factor-ing across national boundries is a way of increasing the number and types of parties that have a vested interest in the profitability of an oil or natural gas project.

Perhaps the most direct way to reduce the company's financial exposure is to raise local capital. Despite the characterization of most developing countries as capital-poor, transnational energy firms often do not have problems in raising large amounts of local debt capital. Local funding gives national entrepreneurs, many of whom may be connected in one way or another with the governing regime, a stake in the profitability of a project. If the government takes actions that curtail profits and cash flow, the effects will be felt by local lenders. Loans in locally denominated cur-rencies, moreover, offer some protection against foreign exchange controls, restric-tions on remittances, and changes in the rate of exchange.

Rather than borrowing locally, a company can pursue a joint venture with private local capital or with the host government. The conventional wisdom is that local equity participation reduces exposure to political risk by "reducing the threat of foreign domination, securing local allies and increasing the role of local business enterprise."[15] This may not always be the case, however. There is some evidence that, although joint ventures with local private capital reduce the likelihood of ex-propriation, ventures with the host government may increase vulnerability to forced divestiture.[16] Similarly, local investors and foreign TNCs are likely to share common interests in the profitability of a project, whereas the host government may be more concerned with policy objectives that conflict with corporate interests. Depending on how the agreement is packaged, moreover, a government with an equity stake in an investment collects revenues in the form of royalties on production and taxes on income as well as direct profits from operations. Because of this multiple sourcing of cash flow, the maximization of the host government's earnings may necessitate a cut in the benefits accruing to the firm-government joint venture.

Joint ventures with private local capital may lend increased legitimacy to foreign firms in the eyes of nationals. Giving the venture a local identity can lessen the degree of nationalism and xenophobia that may arise in response to the presence of giant transnational firms. Similarly, if the host country has a sufficiently developed stock exchange, the firm can invite further involvement on the part of local investors by listing a portion of the share capital on the national exchange.

As with project financing and joint ventures with other transnational firms, ventures with local private interests reduce the equity exposure of a foreign firm. Ownership risks are reduced accordingly, as the company will have fewer assets in the host country that are vulnerable to seizure. Ownership risks can be eliminated totally by the use of service contracts, a form of nonequity investment in which the TNC's direct equity position is zero. Service agreements under which the firm receives a flat fee also insulate the TNC from statutory/administrative and operational risks. Although the company still would be faced with potential transfer risks in repatriating its earnings, it is unlikely that the host government would interfere with such a nonequity relationship. Frequent payment periods, with monthly or at least quarterly payments of fees, reduces the amounts exposed to transfer risks, while making it less worthwhile for the host to interfere with the repatriation of such revenues.

Confronted with a strong probability of expropriation, particularly creeping expropriation, a company can try to stay ahead of the government by corporate-initiated changes.[17] Phased-in local participation, or domestication, of the ownership may satisfy nationalist demands that otherwise might threaten total seizure. The declining value of future cash flows over time and the depletability of any particular oil or gas field makes this an affordable option to the company. The logic of resource economics would seem to be the opposite for the state, which would not be willing to settle for the gradual inheritance of something that is progressively declining in value. This may not be the case, however, since phased-in participation can be employed by the host government as a means of maintaining an acceptable foreign investment regime that satisfies the government's political goals, while allowing for the maturation of the noc and the petroleum-sector-related bureaucracy.

Similarly, a foreign firm can try to stave off or at least delay a political onslaught on its profitability by other tactics. Managing the *pace* of change may be a way of gaining time, during which the company can hope to recover its investment plus an acceptable margin of return. Expanded programs for the training of nationals and the promotion of local managers, for example, can moderate the political pressures on a firm without forcing it to bear undue expenses. The introduction of new managerial procedures or updated technology can be manipulated for the same purpose. Exploitation of the technological edge of the TNCs may seem Machiavellian, but it can be a vital corporate tool in the ongoing bargaining relationship with the host government. This helps explain, in part, popular concerns with transfer of technology issues in the United Nations and Third World-oriented groups.

One of the most standard risk management tactics is to diversify the sources of production or purchases. Minimizing the company's reliance on any particular country is a way of protecting against risks that might arise in that state. Contractual purchases, for example, should be spread among a number of suppliers as well as, perhaps, the spot market to insulate the firm from disturbances in supply or sudden price increases by an exporter. No amount of diversification of sources, however, will protect the company against across-the-board price increases in the international market.

Assets and profits can be insured against politically induced losses. This does not reduce the likelihood of political risk events; rather, it shifts the risk onto the insurer. Government-sponsored insurance with the Overseas Private Investment Corporation (OPIC) may be the exception to the rule, because, in the event of a claim, the host government will be confronted directly by an agency of the U.S. government. This affords both insurance and a variant on the theme of engaging home government support.

OPIC offers a diversified selection of coverages, including inconvertibility, expropriation, war/revolution/insurrection, and interference with operations. These coverages are available on production-sharing agreements, service contracts, and risk contracts, as well as on traditional concession arrangements. OPIC is far too restricted, however, to serve the needs of the oil industry. Coverage is limited to developing countries that have bilateral agreements with the United States and carries a maximum value of $100 million per project and $150 million per country. Moreover, all of the OPEC countries are excluded, and the State Department has the prerogative of terminating the program in countries that persistently violate human rights. Under this latter provision, the Carter administration did not extend OPIC insurance to investments in Argentina, for example.

In addition to OPIC, there is a growing commercial political risk insurance industry that does not have the political constraints under which OPIC operates. Private insurers tend to be even more limited in their financial capacity, however, and they often charge higher premiums than OPIC. Because they are insufficiently served by OPIC and the commercial market, the larger transnational petroleum companies are self-insured. Wholly owned insurance subsidiaries—originally intended for insurance for more traditional types of losses, such as accidents—also are used to provide political risk coverage.

Without interfering in domestic political issues, a company can attempt to mitigate the effect or the causes of political risk events. Returning to models of political risk (see chapter 6), many analysts suggest, for example, that the receipt of U.S. aid reduces the probability of risk events. A politically perspicacious company can act on this understanding in its best interests by lobbying U.S. government officials to give or increase aid to the host country. This might reduce the chance of politically inflicted losses by virtue of the aid relationship and exhibiting to host government leaders that the company is representing the host's interests.

The astute decision maker and political risk analyst must be able to view a project in terms of the costs and benefits to the host state. Although government controls often are seen as costs in the eyes of the TNC, the reliance on foreign firms also carries costs for the host state. The state has its own set of concerns that it wants protected when it enters into an investment agreement. The anticipated benefits of larger, quicker economic returns by relying on foreign corporate know-how, capital, and market outlets must be measured against the costs of foreign management and possible control and the outflow of a share of sales revenues. As one analyst of the host-firm relationship sugggested: "For host country governments, the real issues are how to provide an environment in which transnational corporations can contribute to national goals while minimizing costs and maximizing the benefits [to the host] of private investment."[18]

It is important that firms be sensitive to the cost-benefit position of the host state as a means of protecting corporate interests. Host and firm interests, as discussed in chapter 2, are bound to be different. If a project does not hold the prospect of benefits flowing to the host in excess of the cost that must be borne (however these are measured or calculated), it is unlikely that the government will remain satisfied with the agreement for long. Both in initial negotiations and in subsequent bargaining, the TNC should be certain that the host government gets what its leaders consider fair treatment. Channon and Jallard's political acceptability profile (see chapter 6) intuitively recommends approaching an investment opportunity from the perspective of how a project will satisfy the concerns of the host government. This should keep the firm atuned to the interests of the government and aware of how those interests can be satisfied without disruption of the company-government relationship.

Managing the details of this relationship may be the key to the overall problem of managing political risk. While protecting its interests in an investment, the company needs to be sensitive and responsive to the concerns of the host. This is a sort of dynamic juggling act—the maintenance of a mutually acceptable, ongoing balance. At times, this may be simple, as when both parties have shared interests. At other times, however, the interests of the host and the TNC may be mutually exclusive, prohibiting anything but a winner-takes-all solution. At almost any cost, the company should try to avoid such zero-sum game solutions, as they increase the likelihood of nationalist reaction and expropriation. More typically, however, there is a bundle of interrelated issues about which there is both agreement and disagreement, cooperation and conflict, harmony and competitiveness.

This brings us back to the view of host-firm intercourse as a bargaining relationship. The outcome of any issue or group of issues will be a function of the relative power of the parties and the stakes involved. The mediating variable is negotiating strategy. Timely concessions—such as phased-in participation, as mentioned earlier—may be superior to remaining inflexible, incurring the leadership's wrath, and being expropriated. Almost as important as the outcome of negotiations is the negotiating process itself. The process lends both form and substance to the host-company relationship, insuring a minimal degree of interaction and mutual accommodation.

The negotiating process begins with the conception of interest in an investment and continues throughout the life of a project. Changes in the course of an agreement are inevitable. If changes are to occur—and they will, regardless of corporate acquiescence or opposition—the company has a vested interest that can best be protected by participating in the process of change. Negotiated settlements invariably will be more acceptable to the firm than unilateral government policy declarations. By artful negotiations and the maintenance of an ongoing discussion of host-TNC interests, the company can try to moderate the pace and extent of politically motivated changes and their impact on an investment. At the worst, the company will sincerely represent its interests, will participate in the lowering of the corporate emblem over a project, and will have a word in writing its epitaph. At the best, the firm's management of the negotiating process will be so effective as to short-circuit the impact of any political risk events. In practice, however, one can expect the effective, ongoing management of political risks to yield results for the transnational petroleum or natural gas firm better than those that would have been realized without such management yet not so good as the benefits that would be reaped from an investment free of political risks.

Notes

1. Robert B. Stobaugh, Jr., "How to Analyze Foreign Investment Climates," *Harvard Business Review* 47 (September–October 1969): 101.
2. Melvin A. Conant, "Politics of Oil and Gas," in *Workshop Proceedings: World Oil and Natural Gas Supplies*, ed. H.G. Frank (Palo Alto: Electric Power Research Institute, 1978), p. 3-8-2.
3. Stephen J. Kobrin, "Political Assessment by International Firms: Models or Methodologies," *Journal of Policy Modeling* 3 (1981): 255.
4. Stephen Blank, *Assessing the Political Environment: An Emerging Function in International Companies* (New York: Conference Board, 1980), p. 51.
5. Dan Haendel, *Foreign Investments and the Management of Political Risk* (Boulder: Westview, 1979), p. 135.
6. R.J. Rummel and David A. Heenan, "How Multinationals Analyze Political Risk," *Harvard Business Review* 56 (January–February 1978): 68.
7. Blank, *Assessing the Political Environment*, p. 45.
8. Derek F. Channon and Michael Jallard, *Multinational Strategic Planning*, (New York: American Management Associations, 1978), p. 251.

9. Adeoye A. Akinsanya, *The Expropriation of Multinational Property in the Third World* (New York: Praeger, 1980) p. 286.

10. Ibid., p. 294.

11. Haendel, *Foreign Investments,* p. 143.

12. Sidney M. Leveson, "Effect of the Prospective Investment Climate on Future Oil and Gas Supplies from Foreign Countries," in *Workshop Proceedings: World Oil and Natural Gas Supplies,* ed. H.G. Frank (Palo Alto: Electric Power Research Institute, 1978), p. 3-6-4.

13. Ibid., p. 3-6-5.

14. T.H. Moran, "Transnational Strategies of Protection and Defense by Multinational Corporations: Spreading the Risk and Raising the Cost of Nationalization in Natural Resources," *International Organization* 27 (Spring 1973): 285.

15. Channon and Jallard, *Multinational Strategic Planning,* p. 254.

16. Stephen J. Kobrin, "Foreign Enterprise and Forced Divestment in LDCs," *International Organization* 34 (Winter 1980): 82.

17. Dan Haendel, Gerald T. West, and Robert G. Meadow, eds., *The Measurement of Political Risk and Foreign Investment Strategy* (proceedings) (Philadelphia: Foreign Policy Research Institute, 1975). p. 30.

18. K. Billerbeck and Y. Yasugi, *Private Direct Foreign Investment in Developing Countries,* (World Bank Staff Working Paper No. 348 (Washington, D.C.: World Bank, 1979), p. 3.

Selected Bibliography

Aharoni, Yair. *The Foreign Investment Decision Process*. Cambridge: Harvard University Press, 1966.

Akinsanya, Adeoye A. *The Expropriation of Multinational Property in the Third World*. New York: Praeger, 1980.

Al-Chalabi, Fadhil J. *OPEC and the International Oil Industry: A Changing Structure*. Oxford: Oxford University Press for the Organization of Arab Exporting Countries, 1980.

Al-Otaiba, Mana Saeed. *OPEC and the Petroleum Industry*. New York: Wiley, 1975.

Barrows, Gordon H. "Special Report on World Petroleum Concessions." *Petroleum Economist* 47 (October 1980): 426-428.

Bergsten, C. Fred. "Coming Investment Wars?" *Foreign Affairs* 53 (October 1974): 135-152.

Blank, Stephen. *Assessing the Political Environment: An Emerging Function in International Companies*. New York: Conference Board, 1980.

Boddewyn, Jean, and Cracco, Etienne F. "The Political Game in World Business." *Columbia Journal of World Business* 7 (January-February 1972): 45-56.

Brewer, Thomas L. "Political Risk Assessment for Foreign Direct Investment Decisions: Better Methods for Better Results," *Columbia Journal of World Business* 16 (Spring 1981): 5-12.

Bunn, D.W., and Mustafaoglu, M.M. "Forecasting Political Risk," *Management Science* 24 (November 1978): 1557-1567.

Business International research report. *Strategic Planning for International Corporations*. New York: Business International Corporation, 1979.

Channon, Derek F., and Jallard, Michael. *Multinational Strategic Planning*. New York: American Management Associations, 1978.

Cobbe, James H. *Governments and Mining Companies in Developing Countries*. Boulder: Westview, 1979.

Coplin, William D., and O'Leary, Michael K. "Political Analysis in the Forecast of Oil Price and Production Decisions." Paper presented at the Electric Power Research Institute Fuel Supply Seminar, Memphis, 8-10 December, 1981.

Faber, Mike, and Brown, Roland. "Changing the Rules of the Game: Political Risk, Instability and Fairplay in Mineral Concession Contracts." *Third World Quarterly 2* (January 1980): 100–119.

Fesharaki, Fereidun. "Evolution of Petroleum Agreements: An Economic Interpretation." Paper presented at the East-West Center Workshop, Honolulu, 1980.

Gebelein, C.A.; Pearson, C.F.; and Silbergh, M. "Assessing Political Risk of Oil Investment Ventures." *Journal of Petroleum Technology*, (May 1978), pp. 725–730.

Ghadar, Fariborz. *The Evolution of OPEC Strategy*. Lexington, Mass.: Lexington Books, D.C. Heath, 1977.

Green, Robert T. "Political Structure as a Predictor of Radical Political Change." *Columbia Journal of World Business* 9 (Spring 1974): 28–36.

Haendel, Dan. *Foreign Investments and the Management of Political Risk*. Boulder: Westview, 1979.

Haendel, Dan; West, Gerald T.; and Meadow, Robert G., ed. *The Measurement of Political Risk and Foreign Investment Strategy*, Proceedings of the Conference on the Measurement of Political Risk and Foreign Investment Strategy. Philadelphia: Foreign Policy Research Institute, 1975.

Haendel, Dan; West, Gerald T.; and Meadow, Robert G. *Overseas Investment and Political Risk*. Philadelphia: Foreign Policy Research Institute, 1975.

Haner, F.T. "Rating Investment Risks Abroad." *Business Horizons* 22 (April 1979): 18–23.

Jodice, David A. "Sources of Change in Third World Regimes for Foreign Direct Investment, 1968–1976." *International Organization* 34 (Spring 1980): 177–206.

Johnson, Howard. *Risk in Foreign Business Environments: A Framework for Thought and Management*. Cambridge, Mass.: Arthur D. Little, 1980.

Kapoor, A., and Grub, Phillip D., eds. *The Multinational Enterprise in Transition*. Princeton: Darwin Press, 1972.

Kobrin, Stephen J. "Foreign Enterprise and Forced Divestment in LDCs." *International Organization* 34 (Winter 1980): 65–88.

———. "Political Assessment by International Firms: Models or Methodologies." *Journal of Policy Modeling* 3 (1981): 251–270.

———. "When Does Political Instability Result in Increased Investment Risk?" *Columbia Journal of World Business* 13 (Fall 1978): 113–122.

Lax, Howard L. "Natural Resources in International Politics: Conflict and Cooperation." Paper presented at the CUNY/Political Science Conference, The Graduate School and University Center of the City University of New York, 11–12, December 1981.

Mikdashi, Zuhayr. *The International Politics of Natural Resources*. Ithaca: Cornell University Press, 1976.

Moran, T.H. "Transnational Strategies of Protection and Defense by Multinational Corporations: Spreading the Risk and Raising the Cost for Nationalization in Natural Resources." *International Organization* 27 (Spring 1973): 273–287.

Nagy, P. "Quantifying Country Risk: A System Developed by Economists at the Bank of Montreal." *Columbia Journal of World Business* 13 (Fall 1978): 135–147.

Nankani, Gobind. *Development Problems of Mineral-Exporting Countries*. World Bank Staff Working Paper No. 354. Washington, D.C.: World Bank, 1979.

Prast, W.G., and Lax, Howard L. "Political Risk as a Variable in TNC Decision Making." *Natural Resources Forum* 6 (April 1982): 183–199.

Robock, Stefan H. "Political Risk: Identification and Assessment." *Columbia Journal of World Business* 11 (July–August 1971): 6–20.

Rummel, R.J., and Heenan, David A. "How Multinationals Analyze Political Risk." *Harvard Business Review* 56 (January–February 1978): 67–76.

Rustow, Dankwart A., and Mugno, John F. *OPEC: Success and Prospects*. New York: New York University Press for the Council on Foreign Relations, 1976.

Sigmund, Paul E. *Multinationals in Latin America: The Politics of Nationalization*. Madison: University of Wisconsin Press, 1980.

Smith, David N., and Wells, Louis T., Jr. *Negotiating Third-World Mineral Agreements*. Cambridge, Mass.: Ballinger, 1975.

Stobaugh, Robert B., Jr. "How to Analyze Foreign Investment Climates." *Harvard Business Review* 47 (September–October 1969): 100–108.

Thunell, Lars H. *Political Risks in International Business: Investment Behavior of Multinational Corporations*. New York: Praeger, 1977.

Truitt, Frederick J. *Expropriation of Private Foreign Investment*. Bloomington: Indiana University Press, 1974.

United Nations Commission on Transnational Corporations. *Transnational Corporations in World Development: A Re-Examination*. New York: United Nations Economic and Social Council, 1978.

Valenilla, Luis. *Oil: The Making of a New Economic Order: Venezuelan Oil and OPEC*. New York: McGraw-Hill, 1975.

Wyant, Frank R. *The United States, OPEC and Multinational Oil*. Lexington, Mass.: Lexington Books, D.C. Heath, 1977.

Zorn, Stephen. "Third World Mining Agreements in the 1970s and 1980s." Paper prepared for the United Nations Centre on Transnational Corporations, 1980.

Index